# Noise and
# Vibration Control

# IISc Lecture Notes Series

ISSN: 2010-2402

**Editor-in-Chief:** Gadadhar Misra
**Editors:** Chandrashekar S Jog
  Joy Kuri
  K L Sebastian
  Diptiman Sen
  Sandhya Visweswariah

---

IISc Lecture Notes Series

# Noise and Vibration Control

## M L Munjal
Indian Institute of Science, India

IISc
Press

World Scientific

NEW JERSEY · LONDON · SINGAPORE · BEIJING · SHANGHAI · HONG KONG · TAIPEI · CHENNAI

*Published by*

World Scientific Publishing Co. Pte. Ltd.

5 Toh Tuck Link, Singapore 596224

*USA office:* 27 Warren Street, Suite 401-402, Hackensack, NJ 07601

*UK office:* 57 Shelton Street, Covent Garden, London WC2H 9HE

**British Library Cataloguing-in-Publication Data**
A catalogue record for this book is available from the British Library.

**IISc Lecture Notes Series — Vol. 3**
**NOISE AND VIBRATION CONTROL**

Copyright © 2013 by World Scientific Publishing Co. Pte. Ltd.

ISBN 978-981-4434-73-7

Printed in Singapore

*In pursuit of Quietness*

# Series Preface

## World Scientific Publishing Company - Indian Institute of Science Collaboration

IISc Press and WSPC are co-publishing books authored by world renowned scientists and engineers. This collaboration, started in 2008 during IISc's centenary year under a Memorandum of Understanding between IISc and WSPC, has resulted in the establishment of three Series: IISc Centenary Lectures Series (ICLS), IISc Research Monographs Series (IRMS), and IISc Lecture Notes Series (ILNS).

This pioneering collaboration will contribute significantly in disseminating current Indian scientific advancement worldwide.

The **"IISc Centenary Lectures Series"** will comprise lectures by designated Centenary Lecturers - eminent teachers and researchers from all over the world.

The **"IISc Research Monographs Series"** will comprise state-of-the-art monographs written by experts in specific areas. They will include, but not limited to, the authors' own research work.

The **"IISc Lecture Notes Series"** will consist of books that are reasonably self-contained and can be used either as textbooks or for self-study at the postgraduate level in science and engineering. The books will be based on material that has been class-tested for most part.

Editorial Board for the IISc Lecture Notes Series (ILNS):

Gadadhar Misra, Editor-in-Chief (gm@math.iisc.ernet.in)

Chandrashekar S Jog (jogc@mecheng.iisc.ernet.in)
Joy Kuri (kuri@cedt.iisc.ernet.in)
K L Sebastian (kls@ipc.iisc.ernet.in)
Diptiman Sen (diptiman@cts.iisc.ernet.in)
Sandhya Visweswariah (sandhya@mrdg.iisc.ernet.in)

# Preface

Noise is defined as unwanted sound. Excessive or persistent noise may cause annoyance, speech interference and hearing damage. In working environment, noise may lead to several physiological disorders like high blood pressure, heart problems, headache, etc. Noise is also known to cause accidents at the work place and loss of efficiency and productivity.

Vibration is caused by unbalanced inertial forces and moments. Resonant vibrations may lead to fatigue failures. Flexural vibrations of the exposed surfaces of a machine radiate audible noise, and in fact represents one of the primary sources of noise. Excessive vibration and noise characterize all rotating, reciprocating and flow machinery. This makes automobiles, aeroplanes, thermal power stations, etc. excessively noisy. Thus, the problems of noise and vibration are ubiquitous, cutting across all disciplines of engineering. This book deals primarily with industrial and automotive noise, its measurement and control. The control of noise from vibrating bodies at the source involves control of vibration. Therefore, two of the six chapters deal with vibration, its measurement and control.

It is now well understood that a quieter machine is in every way a better machine. Lesser vibration ensures manufacturing to closer tolerances, lesser wear and tear, and longer fatigue life. Hence, a quieter machine is more cost-effective in the long run. Designing for quietness is known to be most cost-effective. Noise control of existing machinery, while often necessary, calls for stoppage of the machinery and excessive retrofit costs.

The All India Council for Technical Education (AICTE) has listed a course on "Noise and Vibration Control" as a possible elective course for senior undergraduates of the engineering colleges in the country. Such a course would need an appropriate textbook. Hence this presentation.

The author has been teaching this course at the graduate level at the Department of Mechanical Engineering of the Indian Institute of Science (IISc) for over three decades. With a good number of solved as well as unsolved exercises, the present textbook lays stress on design methodologies, applications and exercises. Analytical derivations and techniques are eschewed. Nevertheless, references are provided at the end of each chapter for further study.

This textbook stresses on physical concepts and the application thereof to practical problems. The author's four decades experience in teaching, research and industrial consultancy is reflected in the choice of the solved examples and unsolved problems. The book targets senior undergraduate mechanical engineering students as well as designers of industrial machinery and layouts. It can readily be used for self study by practicing designers and engineers. This is why mathematical derivations have been avoided. The illustrations, tables and empirical formulae have been offered for ready reference.

As Chairman and Member Secretary of the Steering Committee of the Facility for Research in Technical Acoustics (FRITA), Professor D. V. Singh and Mr. S. S. Kohli have played an important role in conceptualizing and supporting this book writing project.

This book has been influenced substantially by Professor Colin H. Hansen whose book 'Engineering Noise Control' I have been following in my graduate course at IISc, and Mr. D. N. Raju with whom I have been collaborating on many of my consultancy projects — in pursuit of quietness. I wish to acknowledge the personal inspiration of Professors Malcolm J. Crocker, M. V. Narasimhan, B. V. A. Rao, S. Narayanan, B. C. Nakra, A. K. Mallik and D. N. Manik, among others. I have drawn heavily from the joint publications of my past as well as present graduate students and research students. My sincere thanks to all of them, particularly, Dr. Prakash T. Thawani and Professor Mohan D. Rao.

I thank Professor R. Narasimhan, Chairman of the Department of Mechanical Engineering, Indian Institute of Science for providing all facilities as well as a conducive environment for research, teaching, consultancy and book writing.

I wish to thank my wife Vandana alias Bhuvnesh for bearing with me during long evenings and weekends that were needed to complete the book.

This text book has been catalyzed and supported by the Department of Science and Technology (DST), under its Utilization of Scientific Expertise of Retired Scientists (USERS) scheme.

Bangalore, April 2013                                         *M. L. Munjal*

# Contents

# Chapter 1

# Noise and Its Measurement

Sound is a longitudinal wave in air, and wave is a traveling disturbance. Mass and elasticity of the air medium are primary characteristics for a wave to travel from the source to the receiver. A wave is characterized by two state variables, namely, pressure and particle velocity. These represent perturbations on the static ambient pressure and the mean flow velocity of wind, respectively. The perturbations depend on time as well as space or distance.

Noise is unwanted sound. It may be unwanted or undesirable because of its loudness or frequency characteristics. Excessive or prolonged exposure to noise may lead to several physiological effects like annoyance, headache, increase in blood pressure, loss of concentration, speech interference, loss of working efficiency, or even accidents in the workplace. Persistent exposure of a worker to loud noise in the workplace may raise his/her threshold of hearing.

The study of generation, propagation and reception of audible sound constitutes the science of Acoustics. There are several branches of acoustics, namely, architectural acoustics, electroacoustics, musical acoustics, underwater acoustics, ultrasonics, physical acoustics, *etc.* The field of industrial noise, automotive noise and environmental noise constitutes engineering acoustics or technical acoustics. This in turn comprises sub-areas like duct acoustics, vibro-acoustics, computational acoustics, *etc.*

The speed at which the longitudinal disturbances travel in air is called sound speed, $c$. It depends on the ambient temperature, pressure and density as follows.

$$c = \left(\gamma RT\right)^{1/2} = \left(\gamma p_0 / \rho_0\right)^{1/2} \tag{1.1}$$

1

Here $\gamma$ is the ratio of specific heats $C_p$ and $C_v$, $R$ is gas constant, $p_0$ is static ambient pressure, $\rho_0$ is mass density, and $T$ is the absolute temperature of the medium. For air at standard pressure $(\gamma = 1.4,\ R = 287.05\ J/(kg.K),\ p_0 = 1.013 \times 10^5\ Pa)$, it can easily be seen that

$$c \simeq 20.05(T)^{1/2} \tag{1.2}$$

where $T$ is the absolute temperature in Kelvin.

Symbol $T$ is used for the time period of harmonic disturbances as well. It is related to frequency f as follows:

$$T = 1/f \text{ or } f = 1/T \tag{1.3}$$

Frequency $f$ is measured in Hertz (Hz) or cycles per second. Wavelength $\lambda$ of moving disturbances, of frequency $f$, is given by

$$\lambda = c/f \tag{1.4}$$

where $c$ denotes speed of sound.

## 1.1  Plane Wave Propagation

Plane waves moving inside a wave guide (a duct with rigid walls) are called one- dimensional waves. These are characterized by the following one-dimensional wave equation [1]:

$$\frac{\partial^2 p}{\partial t^2} - c^2 \frac{\partial^2 p}{\partial z^2} = 0 \tag{1.5}$$

where $p$, $z$ and $t$ are acoustic pressure, coordinate along direction of wave propagation and time, respectively.

For harmonic waves, the time dependence is given by $e^{j\omega t}$ or $\cos(\omega t)$ or $\sin(\omega t)$, where $\omega = 2\pi f$ is the circular frequency in rad/s.

General solution of Eq. (1.5) may be written as

$$p(z,t) = \left(Ae^{-jkz} + Be^{jkz}\right)e^{j\omega t} \tag{1.6}$$

or as

$$p(z,t) = A\, e^{j\omega(t-z/c)} + B e^{j\omega(t+z/c)} \qquad (1.7)$$

where $k = \omega/c = 2\pi/\lambda$ is called the wave number.

It can easily be seen that $A$ is amplitude of the forward progressive wave and B is amplitude of the reflected or rearward progressive wave. Algebraic sum of the two progressive waves moving in opposite directions is called a standing wave. Thus, Eq. (1.6) represents acoustic pressure of a one-dimensional standing wave. The corresponding equation for particle velocity is given by

$$u(z,t) = \frac{1}{\rho_0 c}\left(A e^{-jkz} - B e^{jkz}\right) e^{j\omega t} \qquad (1.8)$$

$\rho_0 c$, product of the mean density and sound speed, represents the characteristic impedance of the medium

For an ambient temperature of $25^0$ C and the standard atmospheric pressure (corresponding to the mean sea level), we have

$p_0 = 1.013 \times 10^5\ Pa,\quad T = 298\ K,\quad \rho_0 = 1.184\ kg/m^3,\quad c = 346\,m/s,$
$\rho_0 c = 410\,kg/(m^2 s)$

One-dimensional wave occurs primarily in the exhaust and tail pipe of automotive engines and reciprocating compressors. These waves are characterized by a plane wave front normal to the axis of the pipe or tube, and therefore they are called plane waves.

The forward wave is generated by the source and the rearward wave is the result of reflection from the passive termination downstream. In particular, $B/A = R$ is called the Reflection Coefficient, and may be determined from the termination impedance [1]. In particular, $R = 0$ for anechoic termination, 1 for rigid (closed) termination, and -1 for expansion into vacuum. In general, $R$ is a function of frequency.

In view of the plane wave character of the one-dimensional waves, the acoustic power flux W of a plane progressive wave may be written as

$$W = IS = \langle p\, u \rangle S = \langle p\, v \rangle,\ v = S\, u \qquad (1.9)$$

where I is Sound Intensity defined as power per unit area in a direction normal to the wave front (in the axial direction for plane waves), $S$ is area of cross-section of the pipe, and $v$ is called the Volume Velocity.

Thus, the acoustic power flux associated with the incident progressive wave and the reflected progressive wave are given by

$$W_i = \frac{|A|^2 S}{2 \rho_0 c} \quad \text{and} \quad W_r = \frac{|B|^2 S}{2 \rho_0 c} \qquad (1.10, 1.11)$$

Note that the factor of 2 in the denominator is due to the mean square values required in the power calculations. Thus, the net power associated with a standing wave is given by

$$W \equiv W_i - W_r = \frac{|A|^2 - |B|^2}{2(\rho_0 c / S)} = \frac{|A|^2 - |B|^2}{2Y} \qquad (1.12)$$

where $Y = \rho_0 c_0 / S$ is the Characteristic Impedance of plane waves, defined as ratio of acoustic pressure and volume velocity of a plane progressive wave along a tube of area of cross-section S.

## 1.2　Spherical Wave Propagation

Wave propagation in free space is characterized by the following Three-Dimensional (3D) Wave Equation [1]

$$\left[ \frac{\partial^2}{\partial t^2} - c^2 \nabla^2 \right] p = 0, \qquad (1.13)$$

where $\nabla^2$ is the Laplacian. In terms of spherical polar coordinates, neglecting angular dependence for spherical waves, Eq. (1.13) can be written as

$$\frac{\partial^2 (rp)}{\partial t^2} - c^2 \frac{\partial^2 (rp)}{\partial r^2} = 0 \qquad (1.14)$$

Comparison of Eqs. (1.5) and (1.14) suggests the following solution for spherical waves:

$$p(r,t) = \frac{1}{r}\left\{ Ae^{-jkr} + Be^{jkr} \right\} e^{j\omega t} \tag{1.15}$$

Substituting it into the momentum equation

$$\rho_0 \frac{\partial u}{\partial t} = -\frac{\partial p}{\partial r} \tag{1.16}$$

yields

$$u(r,t) = \frac{j}{\omega \rho_0 r}\left\{ -\left( jk + \frac{1}{r} \right) Ae^{-jkr} + \left( jk - \frac{1}{r} \right) Be^{jkr} \right\} e^{j\omega t} \tag{1.17}$$

Here $r$ is the radial distance between the receiver and a point source. It may again be noted that the first component of Eqs. (1.15) and (1.17) represents the spherically outgoing or diverging wave and the second one represents the incoming or converging spherical wave. In practice, the second component is hypothetical; in all practical problems dealing with noise radiation from vibrating bodies one deals with the diverging wave only. The ratio of pressure and particle velocity for the diverging progressive wave may be seen to be

$$\frac{p(r,t)}{u(r,t)} = \frac{\omega \rho_0}{k - j/r} = \frac{\rho_0 c}{1 - \dfrac{j}{kr}} = \rho_0 c \frac{jkr}{1 + jkr} \tag{1.18a}$$

It may be observed that unlike for plane progressive waves, this ratio is a function of distance $r$. This indicates that for a spherical diverging wave, pressure and velocity are not in phase. However when the Helmholtz number $kr$ tends to infinity (or is much larger than unity), then this ratio tends to $\rho_0 c$. Physically, it implies that in the far field a spherical diverging wave becomes or behaves as a plane progressive wave. This also indicates that the microphone of the sound level meter should not be near the vibrating surface; it should be in the far field.

In the far field, Helmholtz number is much larger than unity $(kr \gg 1)$ and then Eqs. (1.15), (1.17) and (1.18a) reduce to

$$p(r,t) = \frac{A}{r} e^{-jkr} e^{j\omega t}, \quad u(r,t) = p(r,t)/(\rho_0 c) \qquad \text{(1.18 b, c)}$$

and therefore, sound intensity and total power are given by

$$I(r) = \frac{\text{Re}(p(r)u^*(r))}{2} = \frac{|p(r)|^2}{2\rho_0 c} = \frac{|A|^2}{2\rho_0 c r^2} = \frac{\rho_0 c |u(r)|^2}{2} = \frac{W}{4\pi r^2} \qquad \text{(1.18 d-h)}$$

where $4\pi r^2$ is the surface area of a hypothetical sphere of radius r over which the total power $W$ is divided equally to yield intensity $I(r)$.

**Example 1.1** A bubble-like sphere of 5 mm radius is pulsating harmonically at a frequency of 1000 $Hz$ with amplitude of radial displacement 1 mm in air at mean sea level and $25^0 C$. Evaluate
   (a) amplitude of the radial velocity of the sphere surface;
   (b) amplitude of the acoustic pressure and particle velocity at a radial distance of 1 m.

**Solution**

   (a) For harmonic radial motion, radial velocity u equals $\omega$ times the radial displacement $\xi$, where $\omega = 2\pi f$. Thus $\omega = 2\pi \times 1000$
      $= 6283.2$ rad / s

$$|u| = \omega|\xi| = 6283.2 \times \frac{1}{1000} = 6.283 \text{ m/s}$$

   (b) Wave number, $k = \frac{\omega}{c} = \frac{6283.2}{346} = 18.16 \text{ m}^{-1}$

   Distance, $r = 1$m (given)

   Helmholtz number, $kr = 18.16$ at $r = 1$m, and 0.091 at $r = 0.005$ m (i.e., on the surface). As per Eq. (1.17), for a diverging spherical wave in free field, $B = 0$, and

$$|u|_{surface} = \frac{\left(1+k^2r_0^2\right)^{1/2}}{\omega\rho_0 r_0^2} A$$

or

$$A = \frac{|u|_{surface} \cdot \omega\rho_0 r_0^2}{\left(1+k^2r_0^2\right)^{1/2}} = \frac{6.283 \times 6283.2 \times 1.184 \times (0.005)^2}{\left\{1+(0.091)^2\right\}^{1/2}}$$

$$= \frac{1.1685}{1.004} = 1.1638$$

Substituting this value of $A$ in Eq. (1.17) for $r = 1$m yields

$$|u|_{r=1m} = A \frac{\left(1+k^2r^2\right)^{1/2}}{\omega\rho_0 r^2}\Bigg]_{r=1m} = 1.1638 \times \frac{\left\{1+(18.16)^2\right\}^{1/2}}{6283.2 \times 1.184 \times (1)^2} = 2.845 \times 10^{-3} \text{ m/s}$$

Now, use of Eq. (1.15) yields

$$|p|_{r=1m} = \frac{A}{r} = \frac{1.1638}{1} = 1.164 \text{ Pa}$$

Incidentally, sound pressure amplitude at 1 m may also be obtained by means of Eq. (1.18a):

$$|p|_{r=1m} = |u|_{r=1m} \rho_0 c \frac{18.16}{\left\{1+(18.16)^2\right\}^{1/2}}$$

$$\simeq |u|_{r=1m} \rho_0 c$$

$$= 2.845 \times 10^{-3} \times 410 = 1.166 \text{ Pa}$$

It is worth noting that

$$\frac{p}{u} = \rho_0 c \times jkr \text{ at the surface, where } kr \ll 1,$$

$$\frac{p}{u} = \rho_0 c \text{ in the farfield, where } kr \gg 1,$$

Thus, at the surface (or in the nearfield), sound pressure leads radial velocity by $90^0$ $\left(\text{because } j = e^{j\pi/2}\right)$, whereas in the farfield sound pressure is in phase with particle velocity.

## 1.3 Decibel Level

Human ear is a fantastic transducer. It can pick up pressure fluctuations of the order of $10^{-5}$ Pa *to* $10^3$ Pa; that is, it has a dynamic range of $10^8$ ! Therefore, a linear unit of measurement is ruled out. Instead, universally a logarithmic unit of decibels has been adopted for measurements of Sound Pressure Level, Intensity Level and Power Level. These are defined as follows [1-3]:

$$SPL \equiv L_p = 10 \log \frac{p_{rms}^2}{p_{th}^2} = 20 \log\left(\frac{p_{rms}}{2 \times 10^{-5}}\right), \ dB \qquad (1.19)$$

$$IL \equiv L_I = 10 \log \frac{I}{I_{ref}} = 10 \log \left(\frac{I}{10^{-12}}\right), \ dB \qquad (1.20)$$

$$SWL \equiv L_w = 10 \log \frac{W}{W_{ref}} = 10 \log \left(\frac{W}{10^{-12}}\right), \ dB \qquad (1.21)$$

where *log* denotes log to the base 10, and $p_{th} = 2 \times 10^{-5}$ Pa represents the threshhold of hearing. This standard quantity represents the root-mean-square pressure of the faintest sound of 1000 Hz frequency that a normal human ear can just pick up. The corresponding value of the reference intensity represents

$$I_{ref} = \frac{p_{th}^2}{\rho_0 c} \approx \frac{\left(2 \times 10^{-5}\right)^2}{400} = 10^{-12} \ W/m^2 \qquad (1.22)$$

Similarly, $W_{ref} = 10^{-12}W$.

Making use of Eqs. (1.15) and (1.17) in Eq. (1.20) indicates that in the far field the acoustical intensity is inversely proportional to the radial distance $r$. This inverse square law when interpreted in logarithmic units becomes

$$L_I(2r) - L_I(r) = 10 \ \log \frac{r^2}{(2r)^2} = -6 \ dB \tag{1.23}$$

This indicates that in the far field, the sound pressure level or sound intensity level would decrease by 6 decibels when the measurement distance from the source is doubled.

For spherically diverging waves the sound pressure level at a distance $r$ from a point source is related to the total sound power level as follows [2]:

$$L_p(r) = L_W + 10 \ \log\left(\frac{Q}{4\pi r^2}\right), \ dB \tag{1.24}$$

where Q is the locational directivity factor, given by [2, 3]

$$Q = 2^{n_s} \tag{1.25}$$

Here, $n_s$ is the number of surfaces touching at the source. Thus,
$n_s = 0$ for a source in mid air (or free space),
    1 for a source lying on the floor,
    2 for a source located on the edge of two surfaces, and
    3 for a source located in a corner (where three surfaces meet).
Specifically, for a source located on the floor in the open, Eq. (1.24) yields

$$L_p(r) = L_W - 10 \ \log\left(2\pi r^2\right), \ dB \tag{1.26}$$

## 1.4 Frequency Analysis

The human ear responds to sounds in the frequency range of 20 Hz to 20,000 Hz (20 kHz), although the human speech range is 125 Hz to 8000

Hz. Precisely, male speech lies between 125 Hz to 4000 Hz, and female speech is one octave higher, that is, 250 Hz to 8000 Hz.

The audible frequency range is divided into octave and 1/3-octave bands. For an octave band,

$$f_u / f_l = 2 \text{ and } f_m = \left( f_u . f_l \right)^{1/2} \tag{1.27}$$

so that

$$f_l = \frac{f_m}{2^{1/2}} = 0.707 \, f_m \quad \text{and} \quad f_u = f_m . 2^{1/2} = 1.414 f_m \tag{1.28}$$

Similarly, for a one-third octave band,

$$f_u / f_l = 2^{1/3} \quad \text{and} \quad f_m = \left( f_u . f_l \right)^{1/2} \tag{1.29}$$

so that

$$f_l = \frac{f_m}{2^{1/6}} = 0.891 f_m \quad \text{and} \quad f_u = f_m . 2^{1/6} = 1.1225 f_m \tag{1.30}$$

In Eqs. (1.27) to (1.30), subscripts $l$, u and m denote lower, upper and mean, respectively. It may be noted that three contiguous 1/3-octave bands would have the combined frequency range of the octave band centered at the centre frequency of the middle 1/3-octave band.

1000 Hz has been recognized internationally as the standard reference frequency, and the mid frequencies of all octave bands and 1/3-octave bands have been fixed around this frequency. Table 1.1 gives a comparison of different octave and 1/3-octave bands.

Incidentally, the standard frequency of 1000 Hz happens to be the geometric mean of the human speech frequency range; that is,

$$1000 = \left( 125 \times 8000 \right)^{1/2} \tag{1.31}$$

It may also be noted that

$$2^{1/3} = 1.26 \quad \text{and} \quad 10^{1/10} = 1.259 \tag{1.32}$$

Table 1.1 Bandwidth and geometric mean frequency of standard octave and ⅓-octave bands [12].

| 1 OCTAVE | | | ⅓ OCTAVE | | |
|---|---|---|---|---|---|
| Lower cutoff frequency (Hz) | Center frequency (Hz) | Upper cutoff frequency | Lower cutoff frequency (Hz) | Center frequency (Hz) | Upper cutoff frequency |
| | | | 22.4 | 25 | 28.2 |
| 22 | 31.5 | 44 | 28.2 | 31.5 | 35.5 |
| | | | 35.5 | 40 | 44.7 |
| | | | 44.7 | 50 | 56.2 |
| 44 | 63 | 88 | 56.2 | 63 | 70.8 |
| | | | 70.8 | 80 | 89.1 |
| | | | 89.1 | 100 | 112 |
| 88 | 125 | 177 | 112 | 125 | 141 |
| | | | 141 | 160 | 178 |
| | | | 178 | 200 | 224 |
| 177 | 250 | 355 | 224 | 250 | 282 |
| | | | 282 | 315 | 355 |
| | | | 355 | 400 | 447 |
| 355 | 500 | 710 | 447 | 500 | 562 |
| | | | 562 | 630 | 708 |
| | | | 708 | 800 | 891 |
| 710 | 1000 | 1420 | 891 | 1000 | 1122 |
| | | | 1122 | 1250 | 1413 |
| | | | 1413 | 1600 | 1778 |
| 1420 | 2000 | 2840 | 1778 | 2000 | 2239 |
| | | | 2239 | 2500 | 2818 |
| | | | 2818 | 3150 | 3548 |
| 2840 | 4000 | 5680 | 3548 | 4000 | 4467 |
| | | | 4467 | 5000 | 5623 |
| | | | 5623 | 6300 | 7079 |
| 5680 | 8000 | 11360 | 7079 | 8000 | 8913 |
| | | | 8913 | 10000 | 11220 |
| | | | 11220 | 12500 | 14130 |
| 11360 | 16000 | 22720 | 14130 | 16000 | 17780 |
| | | | 17780 | 20000 | 22390 |

Thus, for practical purposes, $2^{1/3} = 10^{1/10}$. That is why working either way around 1000 Hz, 100 Hz, 200 Hz, 500 Hz, 2000 Hz, 5000 Hz, 10,000 Hz represent the mean frequencies of the respective 1/3-octave bands. It may also be noted that the center frequencies indicated in the $2^{nd}$ and $5^{th}$ columns of Table 1.1 are internationally recognized nominal frequencies and may not be precise.

It may also be noted that the octave band and 1/3-octave band filters are constant percentage band–width filters. The percentage bandwidth of an n-octave filter may be written as

$$bw_n \equiv \frac{f_u - f_l}{f_m} \times 100 = \left[ 2^{n/2} - 2^{-(n/2)} \right] \times 100 \qquad (1.33)$$

Thus, for an octave filter (n = 1), bandwidth is 70.7 % and for a one-third octave filter (n = 1/3), the bandwidth is 23.16% of the mean or centre frequency of the particular filter.

Power spectral density represents acoustic power per unit frequency as a function of frequency. Power in a band of frequencies represents area under the curve within this band. For example, for a flat (constant power) spectrum, the power in a frequency band would be proportional to the bandwidth in Hertz. As the bandwidth of an octave band doubles from one band to the next, the sound pressure level or power level would increase by $10 \log 2 = 3 \text{ dB}$. Similarly, the SPL or SWL of a 1/3-octave band would increase by $10 \log 2^{1/3} = 1 \, dB$, as we move from one band to the next. As a corollary of the phenomenon, SPL in an octave band would be equal to the logarithmic sum of the SPLs of the three contiguous 1/3-octave bands constituting the octave band.

## 1.5  Weighted Sound Pressure Level

The human ear responds differently to sounds of different frequencies. Extensive audiological surveys have resulted in weighting factors for different purposes. Originally, A-weighting was for sound levels below 55 dB, B-weighting was for levels between 55 and 85 dB, and C-weighting was for levels above 85 dB. These are shown in Fig. 1.1.

Significantly, however the A-weighting network is now used exclusively in most measurement standards and the mandatory noise limits.

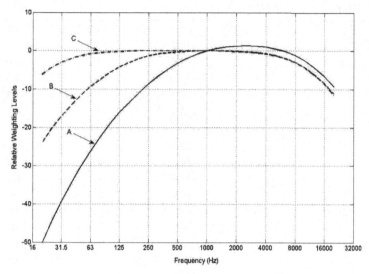

Fig. 1.1 Approximate electrical frequency response of the A-, B-, and C-weighted networks of sound level meters [12].

Table 1.2 contains a listing of the corrections in decibels to be added algebraically to all frequency bands. The A-weighted sound pressure level is denoted as

$$\text{'} L_{pA}, \ dB \text{' or '} L_p, \ dBA \text{'} \tag{1.34}$$

The former notation is more logical. However, the latter continues to be in wide use. Incidentally, symbol dBA is often written as dB(A).

The second column in Table 1.2 indicates octave numbers in popular use among professionals.

## 1.6 Logarithmic Addition, Subtraction and Averaging

The total sound power level of two or more incoherent sources of noise may be calculated as follows:

*Noise and Vibration Control*

Table 1.2  Sound level conversion chart from flat response to A, B, and C weightings.

| Frequency (Hz) | Octave band number | A weighting (dB) | B weighting (dB) | C weighting (dB) |
|---|---|---|---|---|
| 20 | | -50.5 | -24.2 | -6.2 |
| 25 | | -44.7 | -20.4 | -4.4 |
| 31.5 | | -39.4 | -17.1 | -3.0 |
| 40 | | -34.6 | -14.2 | -2.0 |
| 50 | | -30.2 | -11.6 | -1.3 |
| 63 | 1 | -26.2 | -9.3 | -0.8 |
| 80 | | -22.5 | -7.4 | -0.5 |
| 100 | | -19.1 | -5.6 | -0.3 |
| 125 | 2 | -16.1 | -4.2 | -0.2 |
| 160 | | -13.4 | -3.0 | -0.1 |
| 200 | | -10.9 | -2.0 | 0 |
| 250 | 3 | -8.6 | -1.3 | 0 |
| 315 | | -6.6 | -0.8 | 0 |
| 400 | | -4.8 | -0.5 | 0 |
| 500 | 4 | -3.2 | -0.3 | 0 |
| 630 | | -1.9 | -0.1 | 0 |
| 800 | | -0.8 | 0 | 0 |
| 1000 | 5 | 0 | 0 | 0 |
| 1250 | | +0.6 | 0 | 0 |
| 1600 | | +1.0 | 0 | -0.1 |
| 2000 | 6 | +1.2 | -0.1 | -0.2 |
| 2500 | | +1.3 | -0.2 | -0.3 |
| 3150 | | +1.2 | -0.4 | -0.5 |
| 4000 | 7 | +1.0 | -0.7 | -0.8 |
| 5000 | | +0.5 | -1.2 | -1.3 |
| 6300 | | -0.1 | -1.9 | -2.0 |
| 8000 | 8 | -1.1 | -2.9 | -3.0 |
| 10000 | | -2.5 | -4.3 | -4.4 |
| 12500 | | -4.3 | -6.1 | -6.2 |
| 16000 | | -6.6 | -8.4 | -8.5 |
| 20000 | | -9.3 | -11.1 | -11.2 |

$$W_t = \sum_{i=1}^{n} W_i \qquad (1.35)$$

or

$$L_{w,t} = 10 \log\left[\sum_{i=1}^{n} 10^{0.1L_{w,i}}\right]$$ (1.36)

Here n denotes the total number of incoherent sources like machines in a workshop or different sources of noise in an engine, etc. Similarly, the corresponding total SPL at a point is given by

$$L_{p,t} = 10 \log\left[\sum_{i=1}^{n} 10^{0.1L_{p,i}}\right]$$ (1.37)

Incidentally, Eqs. (1.35) – (1.37) would also apply to logarithmic addition of SPL or SWL of different frequency bands in order to calculate the total level. The logarithmic addition of power levels or sound pressure levels have some interesting implications for noise control. It may easily be verified from Eqs. (1.36) and (1.37) that

$$100 \oplus 100 = 103 \ dB$$
$$100 \oplus 90 = 100.4 \ dB$$
$$x \oplus x = x + 3 \ dB$$

Similarly, 10 identical sources of $x$ dB would add upto $x + 10$ dB. Perception wise,

3 dB increase in SPL is hardly noticeable;

5 dB increase in SPL is clearly noticeable; and

10 dB increase in SPL appears to be twice as loud.

Similarly, 10 dB decrease in SPL would appear to be half as loud, indicating 50% reduction in SPL. Therefore, it follows that:

(i) In a complex noisy situation, one must first identify all significant sources, rank them in descending order and plan out a strategy for reducing the noise of the largest source of noise first, and then only tackle other sources in a descending order.

(ii) While designing an industrial layout or the arrangement of machines and processes in a workshop, one must identify and locate the noisiest machines and processes together in one corner, and isolate this area acoustically from the rest of the workshop or factory.

(iii) It is most cost effective to reduce all significant sources of noise down to the same desired level.

Addition of the sound pressure levels of the incoherent sources of noise may be done easily by making use of Fig. 1.2 which is based on the following formula:

$$\Delta L \equiv L_{pt} - L_{p1} = 10 \log\left\{1 + 10^{-0.1\left(L_{p1}-L_{p2}\right)}\right\}, \ dB \qquad (1.38)$$

It may be noted that the addition $\Delta L$ to the higher of the two levels is only 0.4 dB when the difference of the levels ($L_{p1} - L_{p2}$) equals 10 dB. Therefore for all practical purposes, in any addition, if the difference between the two levels is more than 10 decibels, the lower one may be ignored as relatively insignificant.

If one is adding more than two sources, one can still use Fig. 1.2, adding two at a time starting from the lowest, as shown Fig. 1.2

The concept of addition can also be extended to averaging of sound pressure level in a community location. Thus, the equivalent sound pressure level during a time period of 8 hours may be calculated as an average of the hourly readings; that is,

$$L_{p,8h} = 10 \log\left[\frac{1}{8}\sum_{i=1}^{8} 10^{0.1\,L_{p,i}}\right], \ dB \qquad (1.39)$$

This averaging is done automatically in an integrating sound level meter or dosimeter used in the factories in order to ensure that a worker is not subjected to more than 90 dBA of equivalent sound pressure level during an 8 hour shift. Similarly, one can measure $L_d$, the day time average (6 AM to 9 PM) and $L_n$, the night time average (9 PM to 6 AM). Making use of the fact that one needs quieter environment at night, the day-night average (24-hour average) is calculated as follows:

$$L_{dn} = 10 \log\left[\frac{1}{24}\left\{15\times10^{0.1\,L_d} + 9\times10^{0.1\left(L_n+10\right)}\right\}\right], \ dB \qquad (1.40)$$

It may be noted that the $L_n$ has been increased by 10 dBA in order to account for our increased sensitivity to noise at night.

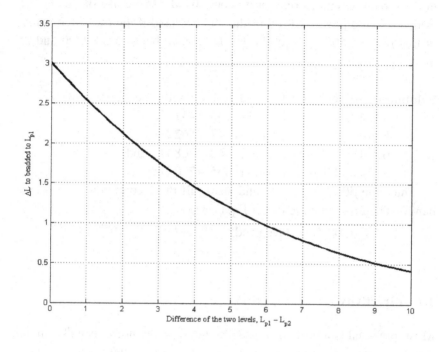

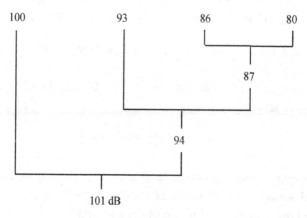

An illustration

Fig. 1.2 Logarithmic addition of two SPLs, $L_{p1} \oplus L_{p2}$.

**Example 1.2** For a reasonably flat frequency spectrum with no sharp peaks or troughs, the band is nearly equal to the sum of the sound powers in the three contiguous one-third octave bands. Make use of this fact to evaluate sound pressure level of the 500 Hz band if the measured values of the 400-Hz, 500-Hz and 630-Hz 1/3-octave bands are 80, 90 and 85 dB, respectively.

**Solution** It may be noted from Table 1.1 that the frequency range of
    400-Hz  1/3-octave band is 355 – 447 Hz,
    500-Hz  1/3-octave band is 447 – 562 Hz,
    630-Hz  1/3-octave band is 562 – 708 Hz, and
    500-Hz  1/1-octave band is 355 – 710 Hz
Thus, the 500-Hz octave band spans all three contiguous 1/3-octave bands. Therefore making use of Eq. (1.37),

$$L_p(500 \ Hz \text{ octave band}) = 10 \log\left(10^{80/10} + 10^{90/10} + 10^{85/10}\right)$$
$$= 91.5 \text{ dB}$$

## 1.7  Directivity

Most practical sources of noise do not radiate noise equally in all directions. This directionality at distance r in the far field is measured in terms of a directivity index DI, or directivity factor DF, as follows:

$$DI_\theta(r) = L_{p,\theta}(r) - L_{p,av}(r) = 10 \log\left(DF_\theta\right), \ dB \qquad (1.41)$$

The average sound pressure level $L_{p,av}$ is calculated from the total sound power level by

$$L_{p,av}(r) = L_W - 10 \log\left(4\pi r^2\right) \qquad (1.42)$$

The average sound pressure level can be evaluated by averaging the measured sound pressure levels at different angles at the same distance (r) around the source, making use of the formula

$$L_{p,av} = 10 \log\left[\frac{1}{n}\sum_{i=1}^{n} 10^{0.1L_{p,i}}\right], dB \qquad (1.43)$$

If all the levels around the machine are within 5 dB of each other, then instead of Eq. (1.43) one can take the arithmetic average of SPLs and add 1 dB to it in order to get a reasonably approximate value of the average sound pressure level.

**Example 1.3** Sound pressure levels at four points around a machine are 85, 88, 92 and 86 dB when the machine is on. The ambient SPL at the four points (when the machine is off) is 82 dB. Calculate the average SPL of the machine alone (by itself).

**Solution**

Making use of Eq. (1.43),

$$L_{p,av}(\text{machine} + \text{ambient}) = 10 \log\left[\frac{1}{4}\left(10^{\frac{85}{10}} + 10^{\frac{88}{10}} + 10^{\frac{92}{10}} + 10^{\frac{86}{10}}\right)\right]$$

$$= 88.6\ dB$$

Incidentally, the arithmetic average of the SPLs at the four points works out to be

$$\frac{1}{4}(85 + 88 + 92 + 86) = 87.7\ dB$$

So, the logarithmic average is quite close 'to the arithmetic average plus 1 dB'.

Now, logarithmically subtracting the ambient SPL we get

$$L_{p,av}(\text{machine alone}) = 10\ log\left(10^{88.6/10} - 10^{82/10}\right)$$

$$= 87.5\ dB$$

## 1.8   Measurement of Sound Power Level

A more precise method of measuring the average sound pressure level
and the total power level of a source consists in making use of an
anechoic room. This is a specially constructed room to simulate free field
environment. Its walls as well as the floor and the ceiling are lined with
long thin wedges of acoustically absorbent material with power
absorption coefficient of 0.99 (99%) or more in the frequency range of
interest. Low frequency noise is not absorbed easily. The lowest
frequency upto which the absorption coefficient is at least 0.99 is called
the cut-off frequency of the anechoic room. Such rooms are used for
precise measurements as per international standards.

However, it is logistically very difficult to mount large and heavy test
machines on suspended net flooring. Therefore the Engineering Method
of measurement of sound pressure level of a machine to a reasonable
accuracy is to make use of a hemi-anechoic room where one simulates a
hemi-spherical free field. This is like testing a machine on the ground in
an open ground. In such a room the floor is acoustically hard (highly
reflective) while all the four walls and the ceiling are lined with highly
absorbent acoustical wedges as indicated above for an anechoic room.

Table 1.3  Co-ordinates of the 12 microphone locations.

| Location No. | $\dfrac{x}{r}$ | $\dfrac{y}{r}$ | z |
|:---:|:---:|:---:|:---:|
| 1 | 1 | 0 | 1.5 m |
| 2 | 0.7 | 0.7 | 1.5 m |
| 3 | 0 | 1 | 1.5 m |
| 4 | -0.7 | 0.7 | 1.5 m |
| 5 | -1 | 0 | 1.5 m |
| 6 | -0.7 | -0.7 | 1.5 m |
| 7 | 0 | -1 | 1.5 m |
| 8 | 0.7 | -0.7 | 1.5 m |
| 9 | 0.65 | 0.27 | 0.71 r |
| 10 | -0.27 | 0.65 | 0.71 r |
| 11 | -0.65 | -0.27 | 0.71 r |
| 12 | 0.27 | 0.65 | 0.71 r |

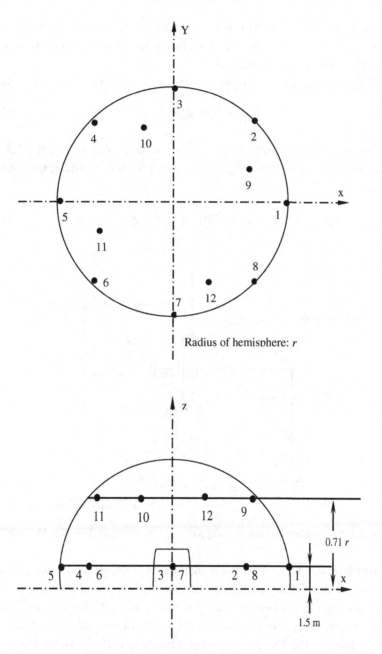

Fig. 1.3 Microphone locations on a hypothetical hemi-spherical surface.

Alternatively, one can use the so-called Survey Method where measurements are made at discrete microphone locations on the parallelepiped shown in Fig. 1.4.

If $S_m$ is the area of the hypothetical measurement surface in $m^2$, then

$$L_w = L_{p,av} + 10 \log(S_m) \qquad (1.44)$$

For a hemi spherical hypothetical surface shown in Fig. 1.3, the measurement surface area $S_m = 2\pi r^2$, and for the parallelepiped surface of Fig. 1.4,

$$S_m = 2a \times 2b + 2(2a+2b)c = 4(ab + bc + ca), \ m^2 \qquad (1.45)$$

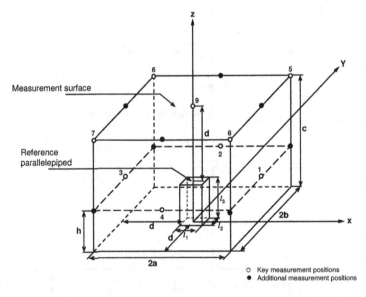

Fig. 1.4 Microphone locations on a hypothetical parallelepiped (Survey Method) [8].

**Example 1.4** The overall dimensions of a diesel generator (DG) set are 3m x 1.5m x 1.2m (height). The A-weighted sound pressure levels at 1 m from the five (4+1) radiating surfaces are 100, 95, 93, 102 and 98 dBA, respectively. Assuming that the contribution from the four walls and ceiling for the DG room is negligible, and making use of the Survey method, evaluate the sound power level of the DG set.

**Solution** With reference to Fig. 1.4,

$$l_1 = 3m, \ l_2 = 1.5m, \ l_3 = 1.2m \ and \ d = 1m$$

Thus,

$$2a = 3 + 2 \times 1 = 5 \ m,$$
$$2b = 1.5 + 2 \times 1 = 3.5 \ m,$$
$$c = 1.2 + 1 = 2.2 \ m$$

Using Eq. (1.45), surface area of the hypothetical measurement surface is given by

$$S_m = 4(ab + bc + ca)$$
$$= 4\left(\frac{5}{2} \times \frac{3.5}{2} + \frac{3.5}{2} \times 2.2 + 2.2 \times \frac{5}{2}\right)$$
$$= 54.9 \ m^2$$

Average value of the SPL may be calculated by means of Eq. (1.43):

$$L_{p,av} = 10 \log\left[\frac{1}{5}\left(10^{100/10} + 10^{95/10} + 10^{93/10} + 10^{102/10} + 10^{98/10}\right)\right]$$
$$= 98.7 \ dBA$$

Finally, the power level of the DG set is given by Eq. (1.44). Thus,

$$L_{W,A} = 98.7 + 10 \ \log(54.9)$$
$$= 116.1 \ dBA$$

## 1.9 Loudness

Loudness index L is measured in terms of sones and the loudness level P in phons. They are related to each other as follows:

$$L = 2^{(P-40)/10}, \quad P = 40 + 33.2 \ \log L \qquad (1.46)$$

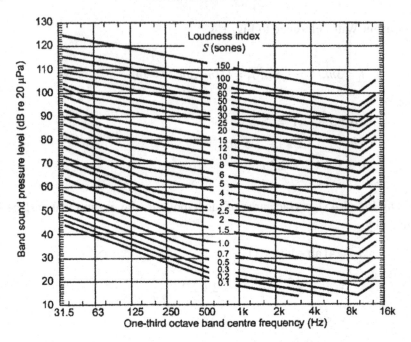

Fig. 1.5 Equal loudness index contours.
(adopted with permission from Bies and Hansen [3]).

The band loudness index for each of the octave bands is read from Fig. 1.5, and then the composite loudness level, $L$ (sones) is determined by [4]

$$L = S_{max} + B\sum_i S_i \qquad (1.47)$$

where $S_i$ is the loudness of the $i^{th}$ band and $S_{max}$ is the maximum of these values. Constant B = 0.3 for octave band analysis and 0.15 for 1/3-octave band analysis. The summation does not include $S_{max}$.

**Example 1.5**   If the measured values of the SPLs for the octave bands with mid-frequencies of 63, 125, 250, 500, 1000, 2000, 4000 and 8000 Hz are 100, 95, 90, 85, 80, 75, 70 and 65 dB, respectively, calculate the total loudness level in sones as well as phons.

**Solution**   The exercise is best done in a tabular form shown below.

Example 1.5 Table

| Octave-band center frequency (Hz) | 63 | 125 | 250 | 500 | 1000 | 2000 | 4000 | 8000 |
|---|---|---|---|---|---|---|---|---|
| Octave-band level (dB) | 100 | 95 | 90 | 85 | 80 | 75 | 70 | 65 |
| Band loudness index (sones) (from Fig. 1.5) | 28 | 25 | 22 | 19 | 17 | 14 | 13 | 11 |

The measured values of SPL are entered in the first row of the table. The band loudness index $S_i$ for each octave band is read from Fig. 1.5 and entered in the second row of the Table. It may be noted that the maximum value of the loudness index, $S_{max}$, is 28 sones. Constant $B = 0.3$ for octave bands, Thus, making use of Eq. (1.47), the composite loudness index, L, is calculated as follows:

$$L = 28 + 0.3\left(25 + 22 + 19 + 17 + 14 + 13 + 11\right)$$
$$= 28 + 0.3 \times 121 = 64.3 \text{ sones}$$

Now use of Eq. (1.46) yields the loudness level in phons:
$$P = 40 + 33.2 \log 64.3$$
$$= 100.0 \text{ phons}$$

## 1.10   Noise Limits in India

The Ministry of Environment and Forests (MOEF) of the Government of India, on the advice of the National Committee for Noise Pollution Control (NCNPC) has been issuing Gazette Notifications prescribing noise limits as well as rules for regulation and control of noise pollution in the urban environment. These are summarized below.

### 1.10.1   *The noise pollution (regulation and control) rules, 2000 [5]*

These rules make use of Table 1.4 for the ambient air quality standards. These are more or less the same as in Europe and USA.

Table 1.4 Ambient air quality standards in respect of noise [7].

| Category of Area/Zone | Limits in Leq (dBA) | |
| --- | --- | --- |
| | Day Time | Night Time |
| Industrial area | 75 | 70 |
| Commercial area | 65 | 55 |
| Residential area | 55 | 45 |
| Silence zone | 50 | 40 |

Note:
1. Day time shall mean from 6:00 a.m. to 10:00 p.m.
2. Night time shall mean from 10:00 p.m. to 6:00 a.m.

(i). A loud speaker or a public address system shall not be used except after obtaining written permission from the authority.

(ii). A loud speaker or a public address system or any sound producing instrument or a musical instrument or a sound amplifier shall not be used at night time except in closed premises for communication within, like auditoria, conference rooms, community halls, banquet halls or during a public emergency.

(iii). The noise level at the boundary of the public place, where loudspeaker or public address system or any other noise source is being used, shall not exceed 10 dB(A) above the ambient noise standards for the area (see Table 1.4) or 75 dB(A) whichever is lower.

(iv). The peripheral noise level of a privately owned sound system or a sound producing instrument shall not, at the boundary of the private place, exceed by more than 5 dB(A) the ambient noise standards specified for the area in which it is used.

(v). No horn shall be used in silence zones or during night time in residential areas except during a public emergency.

(vi). Sound emitting fire crackers shall not be burst in silence zone or during night time.

(vii). Sound emitting construction equipments shall not be used or operated during night time in residential areas and silence zones.

### 1.10.2 *Permissible noise exposure for industrial workers*

In keeping with the practice in most countries, India has adopted the international limit of 90 dBA during an 8-hour shift for industrial workers.

Table 1.5 Permissible noise exposure.

| Duration/day (h) | Sound Level (dBA) Slow response |
|:---:|:---:|
| 16 | 87 |
| 8 | 90 |
| 4 | 93 |
| 2 | 96 |
| 1 | 99 |
| 0.5 | 102 |
| 0.25 or less | 105 |

As shown in Table 1.5, for every 3 dB increase in the A-weighted sound level, the permissible maximum exposure has been reduced to half. Dosimeters have been provided to the factory inspectors and also to the traffic police. The technicians working on noisy machines or in noisy areas are provided with ear muffs or ear plugs, and are required to use them compulsorily.

The total daily dose, D, is given by

$$D = \frac{C_1}{T_1} + \frac{C_2}{T_2} + \ldots\ldots + \frac{C_i}{T_i} + \ldots\ldots + \frac{C_n}{T_n} \qquad (1.48)$$

where

$C_i$ = Total actual time of exposure at a specified noise level, and

$T_i$ = Total time of exposure permitted by the table above at that level.

Alternatively, if we want to evaluate the maximum time that a technician may be asked to work in a noisy environment without risking him to over exposure, we may make use of an integrating sound level meter to evaluate the 8-hour average of the A-weighted SPL as follows:

$$L_{Aeq,8h} = 10 \ \log\left[\frac{1}{8}\int_0^T 10^{L_{pA}(t)/10} \ dt\right] \tag{1.49}$$

where $t$ is in hours. Then, the maximum allowed exposure time to an equivalent *SPL*, $L_{Aeq,8h}$, would be given by

$$T_{allowed} = 8/D \tag{1.50}$$

where D, the daily noise dosage with reference to the base level criterion of 90 dBA is given by

$$D = 2^{(L_{Aeq,8h}-90)/3} \tag{1.51}$$

Here, constant 3 represents the decibel trading level which corresponds to a change in exposure by a factor of two for a constant exposure time (10 log 2 = 3). For use in USA, this constant would be replaced with 5.

**Example 1.6**    The operator of a noisy grinder in an Indian factory is to be protected against over-exposure to the work station noise by rotation of duties. The operator's ear level noise is 93 dBA near the grinder and 87 dBA in an alternative work place. What is the maximum duration during an 8-hour shift that the worker may work on the grinder?

**Solution**

Referring to Table 1.5,
$T_1$ for 93 dBA is 4 hours, and $T_2$ for 87 dBA is 16 hours.
Let the worker operate the grinder for x hours, and work in the quieter location for the remaining duration of 8-x hours.
Use of Eq. (1.48) yields

$$1 = \frac{x}{4} + \frac{8-x}{16}$$

whence

$$x = \frac{8}{3} = 2.67 \text{ hours.}$$

Therefore, the technician should not be made to operate the grinder for more than 2.67 hours during an 8-hour shift.

### 1.10.3  *Noise limit for diesel generator sets*

India has the problem of power scarcity although the government has set up a number of thermal power plants, hydropower plants and atomic power plants.  Therefore, most of the manufacturers have their own captive power plants based on diesel engines. The relevant gazette notification of the MOEF prescribes as follows [6].

* The maximum permissible sound pressure level for new diesel generator (DG) sets with rated capacity upto 1000 KVA shall be 75 dB(A) at 1 metre from the enclosure surface, in free field conditions.
* The diesel generator sets should be provided with integral acoustic enclosure at the manufacturing stage itself.

Noise limits for diesel generator sets of higher capacity shall be as follows [7].

* Noise from DG set shall be controlled by providing an acoustic enclosure or by treating the room acoustically, at the user's end.
* The acoustic enclosure or acoustic treatment of the room shall be designed for minimum 25 dB(A) insertion loss or for meeting the ambient noise standards, whichever is on the higher side.  The measurement for Insertion Loss may be done at different points at 0.5 m from the acoustic enclosure/room, and then averaged.
* The DG set shall be provided with a proper exhaust muffler with insertion loss of minimum 25 dB(A).
* The manufacturer should offer to the user a standard acoustic enclosure of 25 dB(A) insertion loss, and also a suitable exhaust muffler with IL of at least 25 dB(A).

### 1.10.4  *Noise limit for portable gensets*

Most shops and commercial establishments have their kerosene-start, petrol-run portable gensets with power range of 0.5 to 2.5 KVA. The relevant gazette notification for such small portable gensets prescribes as follows [7].

- Sound power level may be determined by means of the Survey method [8] (see Fig. 1.4).
- The A-weighted sound power level of the source in the case of the Direct Method is calculated from the equation

$$L_{WA} = L_{PA} - K + 10 \, \log\left(S/S_0\right) \tag{1.52}$$

where

$K$ is the environmental correction, 10 log (1+4S/A)
$S$ is the area of the hypothetical measurement surface, $m^2$
$S_0 = 1 \, m^2$
$A$ is the room absorption, $m^2$ (see Eq. (4.13) in Chapter 4).

- The prescribed limit of sound power level of portable gensets is 86 dBA.
- This noise limit may necessitate use of acoustic hoods in most cases.

### 1.10.5 *Noise limit for fire crackers*

The Indian Society is a mix of different racial and religious communities. Each community has its festivals that are usually celebrated by means of sound emitting fire crackers. Most of the time, these crackers are fired in hand, and therefore the chance of body damage, particularly to the hearing for the players as well as on-lookers standing nearby, is high. Therefore, the government has mandated as follows [9].

- The manufacture, sale or use of fire-crackers generating noise level exceeding 125 dB(A) or 145 dB(C) peak at 4 meters distance from the point of bursting shall be prohibited.
- For individual fire-cracker constituting a series (joined fire-crackers), the above mentioned limit be reduced by 5 log (N), where N = number of crackers joined together.
- The measurements shall be made on a hard concrete surface of minimum 5 meter diameter or equivalent.
- The measurements shall be made in free field conditions, i.e., there shall not be any reflecting surface upto 15 meter distance from the point of bursting.

## 1.10.6  *Noise limit for vehicles*

Table 1.6  Noise limits for vehicles at manufacturing stage applicable since 1st April 2005 [10].

| Sl. No. | Type of Vehicle | Noise Limits dB(A) |
|---|---|---|
| | Two-wheelers | |
| 1 | Displacement upto 80 cc | 75 |
| 2 | Displacement more than 80 cc but upto 175 cc | 77 |
| 3 | Displacement more than 175 cc | 80 |
| | Three-wheelers | |
| 4 | Displacement upto 175 cc | 77 |
| 5 | Displacement more than 175 cc | 80 |
| | Four-wheelers | |
| 6 | Vehicles used for the carriage of passengers and capable of having not more than nine seats, including the driver's seat | 74 |
| | Vehicles used for carriage of passengers having more than nine seats, including the driver's seat, and a maximum Gross Vehicle Weight (GVW) of more than 3.5 tonnes | |
| 7 | With an engine power less than 150 kW | 78 |
| 8 | With an engine power of 150 kW or above | 80 |
| | Vehicles used for carriage of passengers having more than nine seats, including the driver's seat \ vehicles used for the carriage of goods | |
| 9 | With a maximum GVW not exceeding 2 tonnes | 76 |
| 10 | With a maximum GVW greater than 2 tonnes but not exceeding 3.5 tonnes | 77 |
| | Vehicles used for the transport of goods with a maximum GVW exceeding 3.5 tonnes | |
| 11 | With an engine power less than 75 kW | 77 |
| 12 | With an engine power of 75 kW or above but less than 150 kW | 78 |
| 13 | With an engine power of 150 kW or above | 80 |

Environmental noise of vehicles is measured in a pass-by noise test [10] as shown in Fig. 1.6. The vehicle approaches line A-A at a steady speed corresponding to 3/4 times the maximum power speed of the engine. As the vehicle front end reaches point $C_1$, the accelerator is pushed to full open throttle position and kept so until the rear of the vehicle touches line B-B at point $C_2$. The maximum SPL reading is recorded at the two microphone locations shown in Fig. 1.6. The test is repeated with the vehicle moving in the opposite direction. This is repeated three times. The average of the peak SPL readings represents the pass-by noise of the vehicle. Detailed operating instructions as well as test conditions are given in Ref. [10].

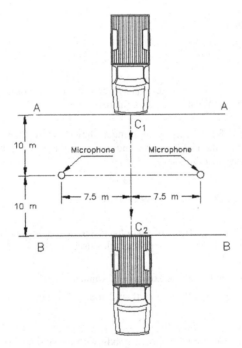

Fig. 1.6 Measurement of passby noise of an automobile [10].

Table 1.6 gives the pass-by noise limits for vehicles at the manufacturing stage [11] applicable since April 2005. These limits are similar to those prescribed in Europe since 1996. The limits are enforced by the Automotive Research Association of India (ARAI, Pune) during the type testing of the new vehicles for establishing their road-worthiness.

## 1.11 Masking

Often environmental noise masks a warning signal. Masking is the phenomenon of one sound interfering with the perception of another sound. This is why honking has to be considerably louder than the general traffic noise around. It has been observed that masking effect of a sound of a particular frequency is more at higher frequencies than at the lower frequencies. Thus if we want to mask ambient sound of 500 Hz to 2000 Hz then we should introduce sound in the 500 Hz one-third octave band. In fact, the masking effect of a narrow band noise is more than that of a pure tone at the centre frequency of the band. The amount of masking is such that a tone which is a few decibels above the masking noise appears to be as loud as it would sound if the masking noise were not present. Masking can be put to effective use in giving acoustic privacy to intellectuals located in open cubicles in large call centers or similar large offices under the same roof. Playing of soft instrumental music in the background is enough to mask the conversation in the neighboring cubicles.

## 1.12 Sound Level Meter

The basic components of a portable sound level meter are microphone, pre-amplifier, weighting network, amplifier, rectifier, the output quantity calculator, and display unit.

The microphone is a transducer for conversion of acoustic signal into a voltage signal. Pre-amplifier is an impedance matching unit with high input impedance and low output impedance. The waiting network relates to A-weighting or C-weighting; B-weighting is rarely used these days. The display unit is essentially a sensitive digital voltmeter, pre-calibrated in terms of sound pressure levels in decibels. For convenience, the attenuators are usually arranged in 10 dB steps. The dynamic range is generally 60 or 80 decibels. There is invariably a meter response function labeled as slow, fast and impulse with averaging time constants of 1 sec, 100 ms and 35 ms, respectively.

The Sound level meter is classified as Class I or Class II (also termed as Type I and Type II), depending on accuracy. Class I sound level meter is a precision sound level meter intended for accurate measurements whereas Class II sound level meter is a general purpose sound level meter intended for field use.

## 1.13 Microphones

There are several types of microphones. Three major types in common use are described here. The most precise microphone consists of a diaphragm which serves as one electrode of condenser and a polarized backing plate separated from it by a very narrow air gap which serves as the other electrode. The condenser is polarized by means of a bound charge so that small variations in the air gap due to pressure induced displacement of the diaphragm result in corresponding variations in the voltage across the condenser. Condenser microphone has relatively flat frequency response, which makes it very desirable for precise measurements.

In a piezo-electric microphone sound incident upon the diaphragm tends to stress or unstress the piezo-electric element which in return induces a bound change across its capacitance. Piezo-electric microphones are also called ceramic microphones.

Table 1.7  Basic microphone types and characteristics.

| Sl. No. | Microphone type | Sensing element | Typical frequency range (Hz) | Temperature and humidity stability |
|---------|-----------------|-----------------|------------------------------|------------------------------------|
| 1 | Condenser | Capacitor | 2-20000 | Fair |
| 2 | Ceramic | Piezoelectric | 20-10000 | Good |
| 3 | dynamic | Magnetic coil | 25-15000 | Good |

Dynamic microphones produce an electrical signal by moving a coil which is connected to a diaphragm through a magnetic field. Obviously, a dynamic microphone must not be used in the vicinity of devices that create magnetic fields; e.g., transformers, motors and alternators. The relative performance of these microphones is given in Table 1.7.

## 1.14 Microphone Sensitivity

Microphone sensitivity is essentially the ratio of electrical output to acoustical input. In logarithmic units it is defined as

$$S = 20 \log\left(\frac{E}{E_{ref}} \quad \frac{p_{ref}}{p}\right), \ dB \tag{1.53}$$

$E_{ref}$, the reference voltage is generally 1 volt and $p_{ref}$ the reference pressure which can be *1 Pa* or *0.1 Pa* (1 microbar). Eq. (1.53) may be rearranged as

$$S = 20 \ \log E - L_p + 94, \ dB \tag{1.54}$$

where $E$ is in volts and $L_p$ is the sound pressure level (re 20 micropascal) on the microphone. Typical values of microphone sensitivities range between −25 and −60dB re 1V/Pa.

Generally there is a trade-off between frequency response and sensitivity. For example, 6-mm diameter microphones have flat frequency response over a wide range of frequencies but they have less sensitivity, as compared to the corresponding 25-mm diameter microphones. Therefore 12-mm diameter microphones are used commonly on most portable sound level meters.

**Example 1.7** A condenser microphone with sensitivity of − 30 *dB re*1.0 V $Pa^{-1}$ is used as the transducer on a portable sound level meter. What would be the SPL reading on the meter if the internal electronic noise is $10 \, \mu V$ ?

## Solution

Use of (1.54) yields

$$-30 = 20 \ \log\left(10 \times 10^{-6}\right) - SPL + 94$$

whence
$$SPL = 30 - 100 + 94$$
$$= 24 \; dB$$

As the internal electronic noise level is 24 dB, and $24 \oplus 30 = 31$, this sound level meter cannot be used to measure SPL of less than 30 dB with an accuracy of ±1 dB.

## 1.15 Intensity Meter

Intensity is defined as the time average of the product of acoustic pressure and normal particle velocity. As per the momentum equation, particle velocity is proportional to pressure gradient. Therefore, an intensity probe consists of two microphones with an acoustically transparent spacer in between. The acoustic pressure is then the average of the pressures picked up by the two microphones individually. Particle velocity is proportional to the difference of the two pressures divided by the distance d between the two (equal to spacer length). Thus,

$$I = \frac{1}{4\omega\rho_0 d} \left( |p_1|^2 - |p_2|^2 \right) \tag{1.55}$$

For this equation to represent the real intensity flux normal to a surface, the intensity probe must be held near the surface in such a way that the axis of the two microphones is perpendicular to the surface. The output of both the microphones is fed to the intensity meter which is calibrated to measure intensity directly in terms of decibels as per Eq. (1.20). Often a good intensity meter has provision for measurement of sound pressure levels and velocity level too. The sound intensity meter is also programmed for the octave and 1/3-octave frequency analysis.

## References

1. Munjal, M. L., Acoustics of Ducts and Mufflers, Wiley, New York, (1987).
2. Irwin, J. D. and Graf, E. R., Industrial Noise and Vibration Control, Prentice Hall, Englewood Cliffs, (1979).

3. Bies, D. A. and Hansen, C. H., Engineering Noise Control, Fourth Edition, Spon Press, London, (2009).
4. Procedure for the Computation of Loudness of Noise, American National Standard USA S3.4-1968, American National Standards Institute, New York, (1968).
5. The Noise Pollution (Regulation and Control) (Amendment) Rules, 2010, MOEF Notification S.O. 50(E), The Gazette of India Extraordinary, (11 January 2010).
6. MOEF Notification G.S.R. 371(E): Environment (Protection) Second Amendment Rules, (2002).
7. MOEF Notification G.S.R. 742(E): Environment (Protection) Amendment Rules, (2000).
8. Acoustics – Determination of sound power levels of noise sources using sound pressure – Survey method using an enveloping measurement surface over a reflecting plane, ISO 3746: 1995(E), International Standards Organization, (1995).
9. MOEF Notification G.S.R. 682(E): Environment (Protection) Second Amendment Rules, (1999).
10. Measurement of noise emitted by moving road vehicles, Bureau of Indian Standards, IS: 3028-1998, New Delhi, (1998).
11. MOEF Notification G.S.R. 849(E): Environment (Protection) Second Amendment Rules, (2002).
12. Joint Departments of the Army, Air Force and Navy, TM 5-805-4/AFJMAN 32-1090, Noise and Vibration Control, (1995).

## Problems in Chapter 1

**Problem 1.1** Often, an approximate value of characteristic impedance $(Z_0 = \rho_0 c)$ of air is taken as a round figure of 400 kg/(m$^2$ s). What temperature does this value correspond to? Adopt the mean sea level pressure.

[Ans.: **$40^0 C$**]

**Problem 1.2** What are the amplitudes of the particle velocity and sound intensity associated with a plane progressive wave of 100 dB sound pressure level? Assume that the medium is air at mean sea level and $25^0$ C.

[Ans.: **6.9 mm/s and 9.76 $mW/m^2$**]

**Problem 1.3** A bubble-like pulsating sphere is radiating sound power of 0.1 Watt at 500 Hz. Calculate the following at a farfield point located 1.0 m away from the centre of the sphere:

    a)    intensity level and sound pressure level
    b)    rms value of acoustic pressure
    c)    rms value of the radial particle velocity
    d)    phase difference between pressure and velocity
        **[Ans.: (a) both 99.0 *dB*, (b) 1.78 Pa, (c) 4.37 mm/s, (d) 6.3⁰]**

**Problem 1.4**    Sound pressure levels at a point near a noisy blower are 115, 110, 105, 100, 95, 90, 85 and 80 dB in the 63, 125, 250, 500, 1000, 2000, 4000 and 8000 Hz octave bands, respectively. Evaluate the total $L_p$ and $L_{pA}$.
        **[Ans.: $L_p$ = 116.6 dB, $L_{pA}$ = 102.4 dBA]**

**Problem 1.5**    Sound pressure levels measured in free space at 12 equispaced locations on a hypothetical hemi-spherical surface of radius 2m around a small portable genset lying on an acoustically hard floor are: 70, 72, 74, 76, 78, 80, 81, 79, 77, 75, 73 and 71 dB, respectively.
    (a).    Evaluate the sound power level of the portable genset.
    (b).    What is the directivity factor for the locations where SPL is maximum and where SPL is minimum?
        **[Ans.: (a) 90.8 dB, (b) 5.25 and 0.525]**

**Problem 1.6**    At a point in a particular environment, the octave-band sound pressure levels are 80, 85, 90, 85, 80, 75, 70 and 65 dB in the octave bands centered at 63, 125, 250, 500, 1000, 2000, 4000 and 8000 Hz, respectively. Evaluate the loudness index in sones for each of these eight octave bands, and thence calculate the total loudness level in sones and phons.
        **[Ans.: 49.6 sones and 96.5 phons]**

**Problem 1.7**    If the hourly A-weighted ambient SPL in an Indian workshop during an 8-hour shift are recorded as 85, 86, 87, 88, 90, 93, 93, 91 dBA, evaluate (a) the daily noise dose of the technicians employed in the workshop, and (b) the maximum permissible exposure time according to the industrial safety standards.

**[Ans.: (a) 1.012 and (b) 7.9 hours]**

**Problem 1.8** The technical specification sheet of a condenser microphone has been misplaced. In order to measure sensitivity of the microphone, it is subjected to SPL of 90 dBA in the 1000-Hz octave band. The output voltage is measured to be 1.0 millivolt. Find the sensitivity of the microphone.

**[Ans.: −56 *dB re* 1V/Pa ]**

# Chapter 2

# Vibration and Its Measurement

Vibration is an oscillatory motion of a particle or an object about its mean position. This motion may or may not be periodic. When vibration is caused by the unbalanced forces or moments of a reciprocating or rotary machine rotating at a constant speed, then it is periodic. In this chapter, and indeed in this textbook, vibration and the resultant acoustic pressure are assumed to be periodic.

Vibration is caused by the interaction of two primary elements: mass and spring (stiffness), or in other words, inertia and elasticity. Damping represents a third element, which is invariably there in the system but is not an essential or primary element for vibration to take place. Mass and spring are characterized by kinetic energy and potential energy, respectively. During a steady state oscillation, there is a continuous exchange between the two types of energy, with their total remaining constant. The external excitation, then, supplies just enough energy to compensate for the energy dissipated into heat by the damping in the system, or radiated out as acoustic energy. In the absence of external excitation, the system would execute free vibrations (if initially disturbed and left alone) with successively decreasing amplitude depending on the amount of damping in the system and radiation resistance on the vibrating surfaces.

In this book, vibration and its control have been dealt with primarily insofar as vibrating structures radiate noise. However, excessive vibration can degrade the performance and decrease the fatigue life of machine bearings and structures, resulting in economic loss. In extreme cases, they may cause fatal accidents.

Study of vibrations involves Newton's laws of motion and hence the science of Dynamics. Vibratory systems are therefore called dynamical systems. These may be classified as linear or nonlinear, lumped or distributed, conservative or nonconservative, depending upon the governing differential equations, and the absence or presence of damping, respectively.

The number of independent coordinates required to describe motion completely is termed the degrees of freedom (DOF) of the system. All basic features of vibration of a multi degree of freedom (MDOF) dynamical system may be easily understood by means of a single degree of freedom (SDOF) system as follows.

## 2.1 Vibration of a Single Degree of Freedom System

For a linear lumped-parameter SDOF system shown in Fig. 2.1, making use of the free body diagram and the Newton's Second Law of Motion, the instantaneous displacement $x(t)$ of the lumped mass m is given by the equation

$$m\,\ddot{x}(t) = f(t) - kx(t) - c\dot{x}(t)$$

or

$$m\,\ddot{x}(t) + c\dot{x}(t) + kx(t) = f(t) \qquad (2.1)$$

where $m$ is the mass, $c$ is the damping coefficient, k is the stiffness of the spring, $f(t)$ is the excitation or external force acting on the mass, $\dot{x}(t) \equiv dx/dt$ is the instantaneous velocity, and $\ddot{x}(t) \equiv d^2x/dt^2$ is the instantaneous acceleration of the mass in the positive (left to right) direction.

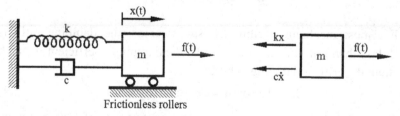

Fig. 2.1 A SDOF system with viscous damping along with the free-body diagram of the mass, m.

Equation (2.1) is a linear, second-order, ordinary differential equation with constant coefficients. Its solution consists of a complementary function and particular integral, representing the free vibration and forced vibration responses of the system, respectively.

### 2.1.1  *Free vibration*

The free vibration response of an underdamped system may be seen to be [1]

$$x(t) = e^{-\zeta \omega_n t} \ (A \sin \omega_d t + B \cos \omega_d t) \qquad (2.2a)$$

where

$\omega_n = (k/m)^{1/2}$ is the natural frequency (in rad/s) of the undamped system (when $c = 0$),

$\omega_d = \omega_n \left(1 - \zeta^2\right)^{1/2}$ is the natural frequency of the damped system,

$\zeta = c / c_c$ is the damping ratio of the system, and

$c_c = 2m\omega_n = 2(km)^{1/2}$ is the critical damping of the system, beyond which there would be no oscillation; the mass would approach the mean position asymptotically.

A and B are arbitrary constants that can easily be determined from the initial displacement $x_0$ and velocity $u_0$ of the mass, $m$. Thus, it can readily be seen that

$$B = x_0 \text{ and } A = \frac{u_0}{\omega_d} + \frac{\zeta \omega_n}{\omega_d} x_0 = \frac{u_0}{\omega_d} + \frac{\zeta}{\left(1 - \zeta^2\right)^{1/2}} x_0 \qquad (2.2b)$$

For most mechanical vibro-acoustic systems, the damping ratio is much less than unity, or in other words, the damping coefficient is much less than the critical damping. Thus,

$$\zeta \ll 1 \quad \text{or} \quad c \ll c_c, \text{ and } \omega_d \simeq \omega_n \qquad (2.3)$$

Figure 2.2 shows the typical free vibration response of such a system, where mass $m$ is given an initial displacement $x_0$, and released with zero initial velocity.

The undamped natural frequency in Hertz, or cycles per second, and the corresponding time period in seconds are given by

$$f_n = \frac{\omega_n}{2\pi}, \quad T = \frac{1}{f_n} = \frac{2\pi}{\omega_n} \tag{2.4}$$

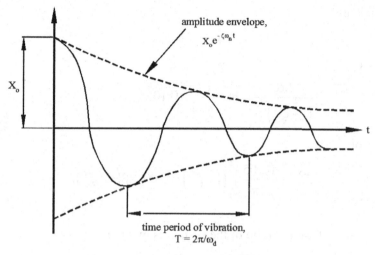

Fig. 2.2  Free vibration of a typical lightly damped system.

**Example 2.1**  The ratio of two successive amplitudes of the damped spring-mass system of the type shown in Fig. 2.1, when disturbed and left to vibrate freely, is 0.9. If the mass is 10 kg and the spring stiffness is 10 N/mm, evaluate the damping ratio, critical damping and the damped natural frequency of the system.

**Solution**

Mass, $m = 10$ kg

Stiffness, $k = 10 \dfrac{N}{mm} \times \dfrac{1000 \; mm}{1 \; m} = 10000$  N/m

Undamped natural frequency of the system,

$$\omega_n = \left( \frac{k}{m} \right)^{1/2} = \left( \frac{10000}{10} \right)^{1/2} = 31.62 \text{ rad/s}$$

As per Eq. (2.2a) and Fig. 2.2, the ratio of successive amplitudes is given by

$$\frac{X_2}{X_1} = e^{-\zeta \omega_n T}, \quad T = 2\pi / \omega_d, \quad \omega_d = \omega_n \left(1 - \zeta^2\right)^{1/2}$$

Thus,

$$e^{-2\pi \zeta / \left(1 - \zeta^2\right)^{1/2}} = 0.9$$

or

$$\frac{2\pi \zeta}{\left(1 - \zeta^2\right)^{1/2}} = \ln\left(\frac{1}{0.9}\right) = 0.105$$

or

$$\zeta^2 = \left(\frac{0.105}{2\pi}\right)^2 \left(1 - \zeta^2\right) = 0.00028\left(1 - \zeta^2\right)$$

Thus, damping ratio of the system is given by

$$\zeta = \left(\frac{0.00028}{1 + 0.0111}\right)^{1/2} = 0.0166$$

Critical damping of the system, $c_c = 2m\omega_n$
$$= 2 \times 10 \times 31.62$$
$$= 632.4 \quad \text{Ns / m}$$

Finally, damped natural frequency of the system,

$$\omega_d = \omega_n \left(1 - \zeta^2\right)^{1/2}$$

$$= 31.62\left\{1 - \left(0.0166\right)^2\right\}^{1/2}$$

$$= 31.62 \times 0.9998 = 31.62 \quad \text{rad / s}$$

$$= \frac{31.62}{2\pi} = 5.03 \text{Hz}$$

This natural frequency and the associated free vibration parameters play a seminal role in our understanding of the forced response of the system, which is of primary concern in the field of noise and vibration control as shown below.

### 2.1.2 *Forced response*

A periodic forcing function with time period T may be expressed as a Fourier series

$$f(t) = a_0 + \sum_{n=1}^{\infty} a_n \cos(n\omega_0 t) + b_n \sin(n\omega_0 t), \quad \omega_0 = \frac{2\pi}{T} \tag{2.5}$$

or as an exponential series

$$f(t) = \sum_{n=-\infty}^{\infty} c_n e^{jn\omega_0 t} \tag{2.6}$$

where

$$a_0 = \frac{1}{T} \int_0^T f(t) dt$$

$$a_n = \frac{2}{T} \int_0^T f(t) \cos(n\omega_0 t) dt$$

$$b_n = \frac{2}{T} \int_0^T f(t) \sin(n\omega_0 t) dt \tag{2.7}$$

$$c_n = (a_n - jb_n)/2 = \frac{1}{T} \int_0^T f(t) e^{-jn\omega_0 t} dt$$

In this book, the exponential form (Eq. 2.6) will be used, with

$$\omega = n\omega_0 \tag{2.8}$$

For a linear system, the principle of superposition holds; i.e., response of a periodic forcing function may be expressed as sum of the system responses to individual harmonics. Thus, Eq. (2.1) for the particular integral may be written as

$$m\ddot{x} + c\dot{x} + kx = F e^{j\omega t} \tag{2.9}$$

Obviously, the steady state response to the harmonic forcing function is given by

$$x(t) = \frac{F\ e^{j\omega t}}{-m\omega^2 + j\omega c + k} \equiv X\ e^{j\omega t} \qquad (2.10)$$

where the complex amplitude of the steady state displacement is given by

$$X = \frac{F}{\left(k - m\omega^2\right) + j\left(\omega c\right)} \qquad (2.11)$$

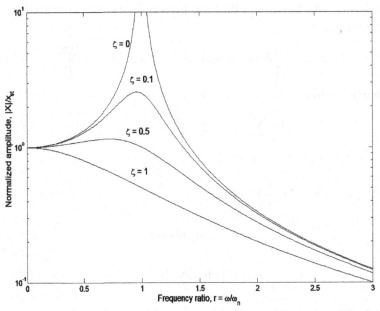

Fig. 2.3 Normalized displacement of mass of the single DOF system of Fig. 2.1 with damping ratio as parameter.

This may be rewritten in the non-dimensional form [2, 3]

$$\frac{X}{F/k} = \frac{1}{1 - \left(\dfrac{\omega}{\omega_n}\right)^2 + j2\zeta\dfrac{\omega}{\omega_n}} \qquad (2.12)$$

or

$$\frac{|X|}{x_{st}} = \frac{1}{\left\{ \left(1 - r^2\right)^2 + \left(2\zeta r\right)^2 \right\}^{1/2}} \quad (2.13)$$

where

$r = \omega / \omega_n$ is the frequency ratio, and

$x_{st} = F / k$ is the static displacement which serves as a normalization factor

The non-dimensional forced response or steady-state response given by Eq. (2.12) is plotted in Fig. 2.3, whence the following observations may be made:

(a) At very low frequencies, $(r \ll 1)$ the dynamic response amplitude $|X|$ tends to the static displacement $x_{st}$.

(b) In the absence of damping, the system response would tend to infinity when the forcing frequency $\omega$ approaches the undamped natural frequency $\omega_n$. This phenomenon is termed Resonance. In fact, the spring of Fig. 2.1 would break in no time if the system were excited at its natural frequency.

(c) At higher frequencies $(\omega \gg \omega_n)$, the dynamic response amplitude decreases monotonically as

$$\frac{|X|}{x_{st}} \simeq \left(\frac{\omega_n}{\omega}\right)^2 \quad (2.13a)$$

Thus, it is imperative to design the system with as low a natural frequency as feasible.

(d) The restraining effect of damping is limited to the resonance frequency and its immediate neighborhood. Nevertheless, its role is crucial.

(e) In a damped single DOF system, resonance peak does not occur precisely at $r = 1$ or $\omega = \omega_n$. As can be seen from Eq. (2.13) and Figure 2.3, it occurs at a slightly lower frequency. However for lightly damped system $(c \ll c_c$ or $\zeta \ll 1)$, this shift can be neglected.

**Example 2.2** A machine of 500 kg mass when lowered onto a set of springs causes a static displacement of 1.0 mm. Evaluate (a) overall stiffness of the springs, and (b) natural frequency of the spring mass system (assume the foundation to be rigid).

**Solution**

Static displacement, $x_{st} = \dfrac{mg}{k}$

whence,

stiffness, $k = \dfrac{mg}{x_{st}} = \dfrac{500 \times 9.81}{1/1000} = 4905 \text{ kN/m}$

Natural frequency, $f_n = \dfrac{\omega_n}{2\pi} = \dfrac{(k/m)^{1/2}}{2\pi} = \dfrac{\left(4905 \times 10^3 / 500\right)^{1/2}}{2\pi} = 15.76$ Hz

## 2.2  Vibration of a Multiple Degrees of Freedom System

A dynamical system with more than one lumped masses interconnected with spring (and damper) elements would require more than one coordinates for description of the instantaneous displacements (and velocities). Such a system is said to have, or be characterized by, multiple degrees of freedom (MDOF). An example of an undamped three DOF system is shown in Fig. 2.4, where, for convenience, the forcing function is assumed to be harmonic.

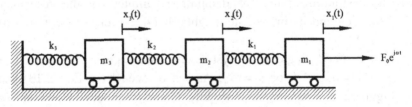

Fig. 2.4  Example of a 3-DOF dynamical system.

Equations of dynamical equilibrium for the system of Fig. 2.4 are given by the matrix equation [1]

$$\begin{bmatrix} m_1 & 0 & 0 \\ 0 & m_2 & 0 \\ 0 & 0 & m_3 \end{bmatrix} \begin{bmatrix} \ddot{x}_1 \\ \ddot{x}_2 \\ \ddot{x}_3 \end{bmatrix} + \begin{bmatrix} k_1 & -k_1 & 0 \\ -k_1 & k_1+k_2 & -k_2 \\ 0 & -k_2 & k_2+k_3 \end{bmatrix} \begin{bmatrix} x_1 \\ x_2 \\ x_3 \end{bmatrix} = \begin{bmatrix} F_0 e^{j\omega t} \\ 0 \\ 0 \end{bmatrix} \quad (2.14)$$

As three coordinates $(x_1, x_2, x_3)$ are sufficient to characterize the system of Fig. 2.4, it is a three degrees of freedom system. Eq. (2.14) may be re-written in the following matrix form:

$$[M]\,\{\ddot{x}\} + [K]\,\{x\} = \{F\}\,e^{j\omega t} \quad (2.15)$$

where

$\{x\}$ is the displacement vector $[x_1,\ x_2,\ x_3]^T$,

$\{\ddot{x}\}$ is the corresponding acceleration vector $[\ddot{x}_1,\ \ddot{x}_2,\ \ddot{x}_3]^T$,

$[M]$, $[K]$ and $\{F\}$ are the inertia (or mass) matrix, stiffness matrix, and force vector, respectively.

## 2.2.1  *Free response*

Free vibration of the system is governed by the equation

$$[M]\,\{\ddot{x}\} + [K]\,\{x\} = \{0\} \quad (2.16)$$

where $\{0\}$ is a null vector $[0\ 0\ 0]^T$, representing free vibration.

Substituting $\{x(t)\} = \{X\}e^{j\omega t}$ in Eq. (2.16), where $\omega$ is a natural frequency, we get a set of three (in general, n) homogeneous equations which can be arranged in the matrix form:

$$\left([K] - \omega^2[M]\right)\{X\} = \{0\}, \quad (2.17)$$

For these equations to be consistent, the determinant of the coefficient matrix must be zero. This condition yields the frequency equation of the system:

$$\left|\ [K] - \omega^2[M]\ \right| = 0 \quad (2.18)$$

Roots of this equation $\left(\omega^2 = \omega_1^2, \omega_2^2, \ldots\ldots\ldots, \omega_n^2\right)$ represent the square of the natural frequencies of the n-DOF system. For the system shown in

Fig. 2.4, n = 3, and therefore this particular system will have three discrete natural frequencies.

## 2.2.2  *Forced response of a multi-DOF system*

Forced response or steady-state response or dynamic response of a multi-DOF system can similarly be evaluated from the inhomogeneous matrix equation (2.15):

$$\{x(t)\} = \left([K] - \omega^2 [M]\right)^{-1} \{F\}\, e^{j\omega t} \qquad (2.19)$$

Here, $\omega$ is the forcing frequency.  It is assumed in Eq. (2.19) that each mass is excited harmonically with the same frequency $\omega$.

Formal similarity of Eq. (2.19) with Eq. (2.10) may be noted. Computationally efficient and inherently stable algorithms, and the corresponding function sub-programs, are available for inversion of the dynamic matrix, [H] in Eq. (2.19) above:

$$[H] \equiv [K] - \omega^2 [M] \qquad (2.20)$$

As indicated by Eqs. (2.14) and (2.20), the inertia matrix [M], stiffness matrix [K], and hence the dynamic matrix [H] are symmetric for linear, passive reciprocal systems. Often, [H] is tridiagonal or banded matrix. Therefore inversion of [H] is a relatively simple and fast operation.

## 2.2.3  *Modal expansion*

When an n-DOF system vibrates at one of its natural frequencies, all of its masses (or lumped inertias) vibrate in phase and the vector of their relative amplitudes represents an eigen vector or modal vector, $\{u\}$. When the modal vectors corresponding to all natural frequencies are arranged column wise, they constitute an n x n modal matrix.

Natural frequencies (or eigenvalues) and the modal matrix (or eigenmatrix) can be computed simultaneously by means of Eigenvalue analysis making use of one of the several algorithms available in Linear

Algebra. Computationally efficient subroutines or Function subprograms are available in the FORTRAN or MATLAB function libraries.

The normal modes or vectors are found to be orthogonal to each other in the following sense [2]:

$$\{u_i\}^T [M]\{u_j\} = \{0\}, \quad \{u_i\}^T [K]\{u_j\} = \{0\} \quad for\ i \neq j \quad (2.21)$$

As a modal vector represents relative amplitudes of different masses during free vibration of the system at one of its natural frequencies, it is desirable to normalize modal vectors such that

$$\{u_i\}^T [M]\{u_i\} = I, \qquad i = 1, 2,\ldots\ldots, n \qquad (2.22)$$

Orthogonality relations (2.21) may be used to decouple the equations of motion of a multi-DOF system. Thus, the solution of the original coupled equations may be reduced to the solution of n independent differential equations. This method of solution is called the Modal Expansion or Eigenfunction Expansion method, and is particularly useful for evaluation of the response of a multi-DOF dynamical system to arbitrary input, $f(t)$. The response or solution may, then, be expressed as a series of the system eigenvectors [2]:

$$\{x(t)\} = \sum_{i=1}^{n} q_i(t) \{u_i\} \qquad (2.23)$$

Here, the coefficients $q_i(t)$ are called Modal Coordinates, and represent a real transformation from the physical coordinates. These may be determined by substituting Eq. (2.23) into the original governing equation

$$[M]\{\ddot{x}\} + [K]\{x\} = \{f(t)\}, \qquad (2.24)$$

and making use of the orthogonality relations (2.21) and normalization relations (2.22). Thus, one obtains the uncoupled or independent differential equations.

$$\{\ddot{q}_i(t)\} + \omega_i^2 \{q_i(t)\} = \{u_i\}^T \{f(t)\}, \quad i = 1, 2,\ldots\ldots, n \qquad (2.25)$$

which can be solved easily. Substituting these modal coordinates in Eq. (2.23), one obtains a closed-form modal expansion of the total response, $x(t)$.

It may be noted that $\{u_i\}^T \{f(t)\}$ on the RHS of Eq. (2.25) represents a modal decomposition of the input force vector inasmuch as it represents the component of the input vector that would excite only the $i^{th}$ mode.

Now, if the forcing vector were harmonic; i.e., if

$$\{f(t)\} = \{F\} e^{j\omega t}, \qquad (2.26)$$

then, the solution of Eq. (2.25) would become

$$q_i(t) = \frac{\{u_i\}^T \{F\}}{\omega_i^2 - \omega^2} e^{j\omega t} \qquad (2.27)$$

Obviously, if the forcing frequency $\omega$ were equal to $\omega_i$, the natural frequency of the $i^{th}$ mode of vibration of the multi-DOF system, then $q_i(t)$, amplitude of the $i^{th}$ mode, would build up until the system broke up, unless the system had substantial inherent damping. This underlines the need for designing-in sufficient damping in every physical dynamical system (like an automobile) that is going to be subjected to a periodic forcing function as well as random excitation. Some practical ways of effecting this are discussed later in Chapter 3.

## 2.3  Transmissibility

The unbalanced forces and moments of a reciprocating machinery (like a reciprocating compressor or engine) or rotating machinery (like a turbine) not only make the machine vibrate but are also transmitted in part to the foundation. The oscillating forces so transmitted propagate through the floor as structure borne sound and radiate audible sound elsewhere. Similarly, oscillation of a surface supporting a sensitive instrument may transmit motion to the instrument. These two phenomena are shown in Fig. 2.5.

Equation of motion of mass m in the single DOF damped system of Fig. 2.5a is given by

$$m\ddot{x} + c\dot{x} + kx = F_0 e^{j\omega t} \qquad (2.28)$$

whence

$$x(t) = \frac{F_0}{k - m\omega^2 + j\omega c} e^{j\omega t} \qquad (2.29)$$

The force transmitted to the foundation is

$$F_T = c\dot{x} + kx = \left( j\omega c + k \right) x(t) = \frac{k + j\omega c}{k - m\omega^2 + j\omega c} F_0 e^{j\omega t} \qquad (2.30)$$

and, then, transmissibility TR is given by

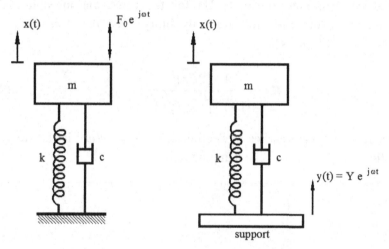

(a) Transmission of force      (b) Transmission of motion

Fig. 2.5  Transmissibility of force and motion.

$$TR \equiv \frac{|F_T|}{F_0} = \frac{\left( k^2 + \omega^2 c^2 \right)^{1/2}}{\left\{ \left( k - m\omega^2 \right)^2 + \omega^2 c^2 \right\}^{1/2}} \qquad (2.31)$$

Now, referring to the single DOF damped system of Fig. 2.5b, the equation of motion is given by

$$m\ddot{x} + c(\dot{x} - \dot{y}) + k(x - y) = 0$$

which can be rearranged as

$$m\ddot{x} + c\dot{x} + kx = c\dot{y} + ky$$

For harmonic excitation $y(t) = Ye^{j\omega t}$, the response will also be harmonic: $x(t) = Xe^{j\omega t}$, and then the motion transmissibility is given by

$$TR \equiv \frac{|X|}{Y} = \left| \frac{j\omega c + k}{-m\omega^2 + j\omega c + k} \right| = \frac{\left( k^2 + \omega^2 c^2 \right)^{1/2}}{\left\{ \left( k - m\omega^2 \right)^2 + \omega^2 c^2 \right\}^{1/2}} \quad (2.32)$$

It may be noted that Eq. (2.32) for the motion transmissibility is identically similar to Eq. (2.31) for the force transmissibility. This common expression for transmissibility has the following non-dimensional form [4]:

$$TR = \left[ \frac{1 + \left( 2\zeta r \right)^2}{\left( 1 - r^2 \right)^2 + \left( 2\zeta r \right)^2} \right]^{1/2} \quad (2.33)$$

where, as defined earlier in Section 2.1.1, $r = \omega/\omega_n$ is the frequency ratio, and $\zeta = c/c_c$ is the damping ratio.

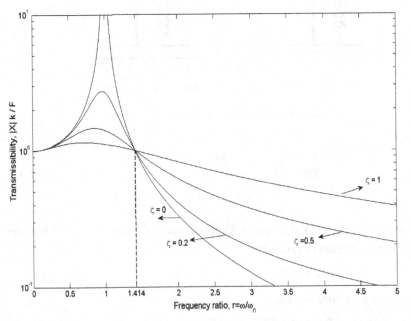

Fig. 2.6 Transmissibility of a single DOF system with damping ratio as a parameter.

Equation (2.33) is plotted in Fig. 2.6. As $r$ tends to unity, TR is inversely proportional to damping ratio $\zeta$. Thus, at around the resonance frequency, in the absence of any damping, TR would tend to infinity. Thus, damping plays a crucial role at and around the resonance frequency $(\omega \approx \omega_n)$. However, it is somewhat counter-productive at $\omega > \sqrt{2}\,\omega_n$, as may be observed from Fig. 2.6

Figure 2.6 also shows that for the force or motion transmissibility to be much less than one, the frequency ratio must be much more than unity. For a lightly damped system with $\omega \gg \omega_n$,

$$TR \approx \frac{\left(1 + 4\zeta^2 r^2\right)^{1/2}}{r^2} \quad for \ \ r \gg 1 \qquad (2.34)$$

This equation may be observed to be similar to Eq. (2.13a) for the response of mass $m$ in Fig. 2.5a:

$$\frac{X}{F_0/k} \approx \frac{1}{r^2} \quad for \ \ r \gg 1 \qquad (2.34a)$$

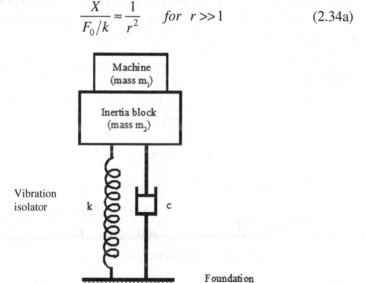

Fig. 2.7 Use of an inertia block.

It follows from Eqs. (2.34) and (2.34a) that in order to reduce the vibration of the machine as well as to reduce the unbalanced forces transmitted to the foundation, the spring stiffness should be as small as feasible so that the natural frequency of the system is much lower than the excitation frequency. This requirement can also be met by means of an inertia block as shown in Fig. 2.7.

Generally, the inertia block mass $m_2$ is several times the machine mass $m_1$. For rigid foundation, natural frequency of the dynamical system of Fig. 2.7 is given by

$$\omega_n = \left( \frac{k}{m_1 + m_2} \right)^{1/2} \qquad (2.35)$$

**Example 2.3** A reciprocating compressor of mass 100 kg is supported on a rigid foundation through a spring of stiffness 1000 N/cm in parallel with a dashpot of damping coefficient 100 Ns/m. If it is acted upon by a vertical unbalanced force of 1000 N amplitude at a frequency corresponding to 1500 RPM, then evaluate amplitude of (a) vibration of the compressor body, and (b) the force transmitted to the rigid foundation.

**Solution**

Mass, $m = 100$ kg  (*given*)

Stiffness, $k = 1000 \dfrac{N}{cm} = 1000 \dfrac{N}{0.01m} = 10^5 \, N/m$

Damping coefficient, $c = 100$  Ns/m  (*given*)

Forcing frequency, $\omega = \dfrac{2\pi RPM}{60} = \dfrac{2\pi \times 1500}{60} = 157.1$  rad/s

Undamped natural frequency,

$$\omega_n = \left( \frac{k}{m} \right)^{1/2} = \left( \frac{10^5}{100} \right)^{1/2} = 31.62 \text{ rad/s}$$

Frequency ratio, $r = \dfrac{\omega}{\omega_n} = \dfrac{157.1}{31.62} = 4.97$

Amplitude of the excitation force, $F_0 = 1000$ N  (given)

(a) Amplitude of vibration of the compressor body is given by Eq. (2.29):

$$X = \frac{F_0}{\left\{\left(k - m\omega^2\right)^2 + \omega^2 c^2\right\}^{1/2}}$$

$$= \frac{1000}{\left[\left\{10^5 - 100 \times (157.1)^2\right\}^2 + (157.1 \times 100)^2\right]^{1/2}}$$

$$= \frac{1000}{2.37 \times 10^6} = 0.42 \text{ mm}$$

(b) Amplitude of the unbalanced force transmitted to the rigid foundation is given by Eq. (2.30):

$$F_T = \frac{\left(k^2 + \omega^2 c^2\right)^{1/2}}{\left\{\left(k - m\omega^2\right)^2 + \omega^2 c^2\right\}^{1/2}} F_0$$

$$= \frac{\left\{\left(10^5\right)^2 + (157.1 \times 100)^2\right\}^{1/2} \times 1000}{\left[\left\{10^5 - 100(157.1)^2\right\}^2 + (157.1 \times 100)^2\right]^{1/2}}$$

$$= \frac{1.012 \times 10^8}{2.37 \times 10^6} = 42.7 \text{ N}$$

Incidentally, force transmissibility,

$$TR = \frac{F_T}{F_0} = \frac{42.7}{1000} = 0.0427 = 4.27\%$$

It is worth noting that the vibration isolation system of Example 2.3 is very efficient inasmuch as amplitude of vibration of the compressor body is as small as 0.42 mm and the force transmissibility is a meager 4.27%. This effectiveness is due to the fact that the natural frequency is so small that frequency ratio,

$$r = \frac{\omega}{\omega_n} = \frac{157.1}{31.62} = 4.97 \, ,$$

which is much more than unity. This is indeed the basic principle underlying the design of vibration isolator of a single DOF system.

## 2.4   Critical Speed

Critical speed is the angular speed of rotation of a rotor which coincides with the flexural natural frequency of the rotor-shaft-bearings system.

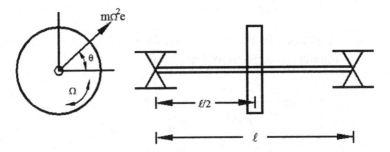

Fig. 2.8  A rotor and shaft mounted on simple support bearings.

Let a rotor or disc of mass $m$ be mounted in the middle of a shaft of length $l$, as shown in Fig. 2.8. There would invariably be a little unbalance (eccentricity of the centre of gravity) in the rotor, resulting in a centrifugal force $m \, \Omega^2 \, e$, where $e$ is the eccentricity and $\Omega$ is the angular speed of rotation of the system. It will have a vertical component $m \, \Omega^2 \, e \, sin\theta$ where $\theta = \Omega \, t$. This will result in a vertical oscillation of the rotor, $y(t)$, governed by the equation:

$$m\ddot{y} + k_f \, y = m \, \Omega^2 \, e \, \sin(\Omega t) \tag{2.36}$$

where $k_f$ is the flexural stiffness of the simply supported shaft [5], $k_f = 48EI/l^3$ ,

EI is the flexural rigidity of the shaft. Here, mass of the shaft and damping of the bearings have been neglected for simplicity.

Equation (2.36) has the following steady-state solution:

$$y(t) = \frac{m \, \Omega^2 e}{k_f - m \, \Omega^2} \, \sin(\Omega t) \tag{2.37}$$

It can be rearranged in the non-dimensional form

$$y(t) = \frac{r_f^2}{1 - r_f^2} \, e.\sin(\Omega \, t) \tag{2.38}$$

where $r_f = \dfrac{\Omega}{\omega_{n,f}}$ is the flexural frequency ratio

$\omega_{n,f} = \left(k_f / m\right)^{1/2}$ is the flexural natural frequency for the simplified single DOF system

Clearly, $y(t)$ would tend to infinity if $r_f$ were to be unity, or if $\Omega = \omega_{n,f}$. Therefore, the critical speed of rotation of the rotor is $\omega_{n,f}$ rad/s or $\left(\omega_{n,f} / 2\pi\right) \times 60$ revolutions per minute. Hence, the critical speed $N_c$ is given by

$$N_c = \frac{60}{2\pi} \left( \frac{48EI}{ml^3} \right)^{1/2}, \quad RPM \tag{2.39}$$

This is of course an approximate estimate of the critical speed of the simplified or idealized undamped single DOF rotor shown in Fig. 2.8. Nevertheless, it explains the concept of critical speed.

**Example 2.4** A steel disc of diameter 0.5 m and width 5 cm is mounted in the middle of a 2 cm diameter, 1 m long shaft mounted on ball bearings at the two ends. Evaluate the critical speed of the rotary system.

**Solution**

Diameter of the rotary disc, $D = 0.5$ m (*given*)

Axial width (or length) of the disc, $b = 0.05$ m (*given*)

For steel, elastic modulus, $E = 2 \times 10^{11}$ N / m$^2$

density, $\rho = 7800$ kg / m$^3$

Mass of the disc, $m = \rho \dfrac{\pi D^2}{4} b$

$$= \frac{7800 \times \pi (0.5)^2 \times 0.05}{4}$$

$$= 76.58 \text{ kg}$$

Diameter of the shaft, $d = 0.02$ m  (*given*)

Moment of inertia of the shaft, $I = \dfrac{\pi d^4}{64} = \dfrac{\pi \times (0.02)^4}{64}$

$$= 7.854 \times 10^{-9} \text{ m}^4$$

Ball bearings would behave as simple supports for the shaft.
Length of the simply supported shaft (see Fig. 2.8), $l = 1$ m (given)
Finally, the critical speed of the system of Fig. 2.8 is given by Eq. (2.39):

$$N_c = \frac{60}{2\pi} \left( \frac{48\, EI}{ml^3} \right)^{1/2} = \frac{60}{2\pi} \left( \frac{48 \times 2 \times 10^{11} \times 7.854 \times 10^{-9}}{76.58 \times (1)^3} \right)^{1/2}$$

$$= 299.6 \text{ RPM}$$

## 2.5  Dynamical Analogies

The electro-mechanical analogies and electro-acoustic analogies constitute a general class of dynamical analogies. These analogies are symbolic as well as physical (conceptual). In other words, these analogies imply similarity of the governing equations as well as physical behaviour of the corresponding dynamical elements and state variables. Electromotive force (or voltage) and current in the electrical networks correspond to mechanical force and velocity in the mechanical vibrational systems. Similarly, electrical inductance, capacitance and resistance are analogous to inertia (or mass), spring compliance (inverse of stiffness) and damping, respectively [6]. Equations of dynamical equilibrium are found to be similar to loop equations in the analogous electrical networks, as illustrated in Fig. 2.9 for a two-DOF system.

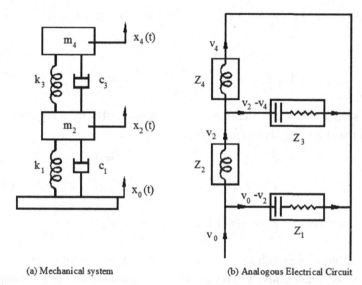

(a) Mechanical system                    (b) Analogous Electrical Circuit

Fig. 2.9 An illustration of the electro-mechanical analogies: a system excited at the bottom support (constant velocity source or infinite-impedance source).

The free body diagrams of the lumped masses $m_2$ and $m_4$ yield the following equations of dynamical equilibrium:

$$m_2 \ddot{x}_2 + c_1 (\dot{x}_2 - \dot{x}_0) + k_1 (x_2 - x_0) + c_3 (\dot{x}_2 - \dot{x}_4) + k_3 (x_2 - x_4) = 0 \quad (2.40)$$

$$m_4 \ddot{x}_4 + c_3 (\dot{x}_4 - \dot{x}_2) + k_3 (x_4 - x_2) = 0 \quad (2.41)$$

For harmonic excitation (time dependence $e^{j\omega t}$), the time derivative $d/dt$ is equivalent to a multiplication factor $j\omega$, and displacement $x(t)$ and accelerations $\ddot{x}(t)$ are related to velocity $v = \dot{x}(t)$ as follows:

$$\ddot{x} = \frac{d^2 x}{dt^2} = \frac{dv}{dt} = j\omega v \quad (2.42)$$

Similarly,

$$x = \frac{v}{j\omega} \quad (2.43)$$

Thus, the ordinary differential Eqs. (2.40) and (2.41) reduce to the following algebraic equations:

$$\left( j\omega\, m_2 \right) v_2 + c_1 \left( v_2 - v_0 \right) + \frac{k_1}{j\omega}\left( v_2 - v_0 \right) + c_3 \left( v_2 - v_4 \right) + \frac{k_3}{j\omega}\left( v_2 - v_4 \right) = 0$$

$$\text{(2.44)}$$

$$\left( j\omega\, m_4 \right) v_4 + c_3 \left( v_4 - v_2 \right) + \frac{k_3}{j\omega}\left( v_4 - v_2 \right) = 0 \qquad \text{(2.45)}$$

These equations are analogous to the Kirchhoff's loop equations for the electrical analogous circuit of Fig. 2.9 (b):

$$Z_2\, v_2 + Z_1 \left( v_2 - v_o \right) + Z_3 \left( v_2 - v_4 \right) = 0 \qquad \text{(2.46)}$$

$$Z_4\, v_4 + Z_3 \left( v_4 - v_2 \right) = 0 \qquad \text{(2.47)}$$

$$Z_2 = j\omega\, m_2, \; Z_4 = j\omega\, m_4, \; Z_1 = c_1 + \frac{k_1}{j\omega}, \; Z_3 = c_3 + \frac{k_3}{j\omega} \qquad \text{(2.48)}$$

Equations (2.48) indicate that mass is analogous to inductance, spring stiffness is analogous to reciprocal of capacitance, and damping coefficient is analogous to resistance, respectively, in the electrical networks.

It may be noted from Fig. 2.9 that the free end of mass $m_4$ in Fig. 2.9a is represented as a zero force or zero impedance or a short circuit in Fig. 2.9b. As the force across a spring is proportional to differential displacement at its ends, it is represented as a shunt element. Similarly, as the force across a damper is proportional to differential velocity at its ends, it is also represented as a shunt element. However, the spring and damper that are parallel to each other and share the same terminations or ends, are in series with each other within the same shunt impedance in Fig. 2.9b (see $Z_1$ and $Z_3$).

Obviously dynamical analogies illustrated in Fig. 2.9 provide an alternative analytical tool for evaluation of forced or steady-state response of multi- as well as single-DOF dynamical systems. They also help in deriving the natural frequency equation of a freely vibrating system.

An important advantage of dynamical analogies is that they help in better conceptualization or understanding of the physical function of different dynamical elements. This in turn helps in synthesis of vibration absorbers as well as isolators for different applications, as will become clear in the next chapter.

## 2.6   Vibration of Beams and Plates

Noise emitted by vibrating bodies (mostly thin plates and sheet-metal components) is one of the major sources of industrial as well as automotive noise. Control of this type of noise requires knowledge of the free as well as forced vibration of plates and plate-like surfaces. The equations governing the flexural vibration of plates may be conceptualized and developed as two-dimensional (2-D) extension of the corresponding equations for transverse vibration of uniform beams where the effects of rotary inertia and shear deformation are neglected.

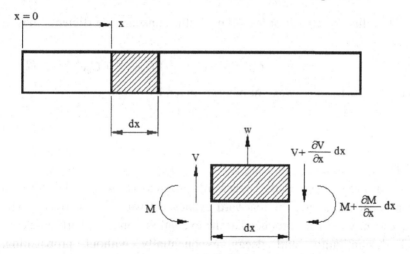

Fig. 2.10  Positive directions of the transverse displacement w,
bending moment M and shear force V.

For a freely vibrating beam, the loading on the beam (see Fig. 2.10 for nomenclature and positive directions of the transverse displacement *w*, shear force V and bending moment M) will be equal to the inertia force. Thus,

$$EI \frac{\partial^4 w}{\partial x^4} = -\rho A \frac{\partial^2 w}{\partial t^2} \qquad (2.49)$$

where $E$ is the Young's modulus of the beam material, I is the second moment of area of the cross-section about the neutral plane axis of the beam, $\rho$ is the density and $A$ is the cross-sectional area of the beam. Thus $\rho A$ represents the mass per unit length of the beam. The product $EI$ is often referred to as flexural rigidity or flexural stiffness of the beam.

For harmonic time dependence,

$$w(x,t) = w(x)e^{j\omega t}, \qquad (2.50)$$

Equation (2.49) reduces to the ordinary differential equation (ODE):

$$EI \frac{d^4 w(x)}{dx^4} - \rho A \omega^2 w(x) = 0 \qquad (2.51)$$

This linear, fourth-order ODE with constant coefficients has a general solution

$$w(x) = C_1 e^{-jk_b x} + C_2 e^{jk_b x} + C_3 e^{-k_b x} + C_4 e^{k_b x} \qquad (2.52)$$

where $k_b$, the bending wave number, is given by

$$k_b = \left( \frac{\rho A \omega^2}{EI} \right)^{1/4} \qquad (2.53)$$

The four terms constituting Eq. (2.52) may be shown to represent the forward progressive wave, rearward progressive wave, the forward evanescent wave, and the rearward evanescent wave, respectively. The evanescent waves are produced at the excitation point, support, boundary and discontinuity, and decay exponentially without propagating. Therefore, the evanescent terms are not waves in a rigorous sense inasmuch as wave is defined as a moving disturbance. The evanescent terms are near-field effects.

Comparing the bending wave number $k_b$ given by Eq. (2.53) with the acoustic wave number $k = \omega/c$ (see Eq. 1.6), the bending wave speed is given by [1]

$$c_b = \frac{\omega}{k_b} = \left(\frac{EI\omega^2}{\rho A}\right)^{1/4} \quad (2.54)$$

It may be noted that $c_b$ is a function of frequency, unlike the sound speed c which is independent of frequency. Thus, different harmonics of a flexural disturbance will move at different speeds along the beam. This phenomenon is called Dispersion. In other words, unlike sound waves, the flexural or bending waves are dispersive.

Constants $C_1 - C_4$ and wave number $k_b$ (and thence the natural frequency $\omega$) may be determined from the boundary conditions. For example, for the cantilever beam shown in Fig. 2.11, the boundary conditions are as follows.

At the fixed (clamped) end, $x = 0$:

$$\text{displacement } w(0) = 0 \text{ and slope } \frac{dw}{dx}(0) = 0 \quad (2.55)$$

At the free end, $x = l$:

$$\text{bending moment } EI\,\frac{d^2 w(l)}{dx^2} = 0 \text{ and shear force } EI\,\frac{d^3 w(l)}{dx^3} = 0 \quad (2.56)$$

Thus, it can be shown that the frequency equation is a transcendental equation [1]:

$$\cos(k_b l)\,\cosh(k_b l) = -1 \quad (2.57)$$

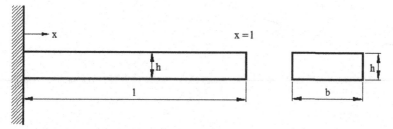

Fig. 2.11 Flexural vibration of a uniform cantilever beam of length $l$, width b and thickness $h$.

This has infinite number of roots, each of which corresponds to a natural frequency and a corresponding mode shape. Thus, displacement

$w(x)e^{j\omega t}$ in response to an external transverse force $F(x)e^{j\omega t}$ may be expressed as a sum of all individual modes:

$$w(x) = \sum_{i=1}^{\infty} A_i \ \phi_i(x) \qquad\qquad (2.58)$$

where $\phi_i$, the $i^{th}$ mode shape of the cantilever of Fig. 2.11, is given by [1,7]

$$\phi_i(x) = \cosh\left(k_{bi}x\right) - \cos\left(k_{bi}x\right) - \left\{ \frac{\cosh(k_{bi}l) + \cos(k_{bi}l)}{\sinh(k_{bi}l) + \sin(k_{bi}l)} \right\} \left\{ \sinh(k_{bi}x) - \sin(k_{bi}x) \right\}$$

$$(2.59)$$

Coefficient $A_i$, representing the relative strength or amplitude of the $i^{th}$ mode, may be found by making use of the orthogonality of natural modes.

In practice, the natural frequencies, modal shapes and response of a beam to a forcing function are found numerically on a digital computer, making use of the Finite Element Model (FEM) and/or measurements.

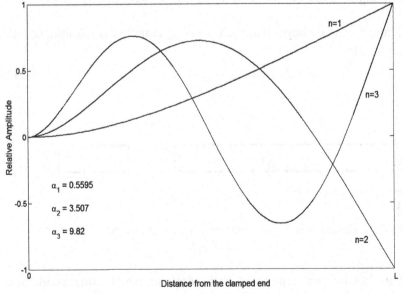

Fig. 2.12 The first three normal modes of vibration of the cantilever beam of Fig. 2.11.

Figure 2.12 shows the first three bending mode shapes for the cantilever of Fig. 2.11. The corresponding natural bending wave numbers are also indicated therein. The $i^{th}$ natural frequency can be found by means of Eq. (2.53). Thus,

$$f_{bi} = \frac{k_{bi}^2}{2\pi}\left(\frac{EI}{\rho A}\right)^{1/2} = \alpha_i \left(\frac{EI}{\rho A l^4}\right)^{1/2} \qquad (2.60)$$

where $\alpha_1 = 0.5595$, $\alpha_2 = 3.507$ and $\alpha_3 = 9.82$ for the first three natural modes of the cantilever vibration shown in Fig. 2.11 [1].

Relations (2.49) to (2.60) for flexural vibration of a beam find their counterparts in vibration of thin rectangular plates as follows.

The two-dimensional wave equation for flexural or transverse vibration of a thin rectangular plate is:

$$\frac{Eh^3}{12\left(1-v^2\right)}\left\{\frac{\partial^4 w}{\partial x^4}+2\frac{\partial^4 w}{\partial x^2 \partial y^2}+\frac{\partial^4 w}{\partial y^4}\right\}+\rho h\,\frac{\partial^2 w}{\partial t^2}=0 \qquad (2.61)$$

where h is the plate thickness (assumed to be uniform), $v$ is Poisson's ratio, $\rho$ is the density, so that the product $\rho h$ represents the mass per unit area (also called surface density) of the plate, and $w = w(x, y, t)$ is the transverse plate displacement.

Assuming time dependence to be $e^{j\omega t}$ as before, Eq. (2.61) reduces to a linear, fourth-order ODE, which may be solved by means of separation of variables. Then, making use of the appropriate boundary conditions for a rectangular plate of dimensions $l_x$ and $l_y$, one gets the following relationships [1, 7]:

$$c_{Lp} = \left\{\frac{E}{\rho\left(1-v^2\right)}\right\}^{1/2} \qquad (2.62)$$

$$c_{bp} = \left\{\frac{Eh^3\omega^2}{12\left(1-v^2\right)\rho h}\right\} = \left(1.8c_{Lp}hf\right)^{1/2} \qquad (2.63)$$

$$k_{bp} = \left(k_x^2 + k_y^2\right)^{1/2} \qquad (2.64)$$

Here, $c_{Lp}$ is the quasi-longitudinal wave velocity,

$c_{bp}$  is the bending wave velocity for  thin plate,

$k_x$ *and* $k_y$ are the *x*-component and *y*-component of  $k_{bp}$ , the
bending wave number (considered a vector).

For a simply supported plate,

$$k_x = \frac{m\pi}{l_x} \text{ for } m = 1,2,3,\ldots\ldots\ldots \tag{2.65}$$

$$k_y = \frac{n\pi}{l_y} \text{ for } n = 1,2,3,\ldots\ldots \tag{2.66}$$

$$f_{m,n} = \left(1.8\ c_{Lp}h\right)\left\{\left(\frac{m}{2l_x}\right)^2 + \left(\frac{n}{2l_y}\right)^2\right\}, Hz \tag{2.67}$$

where *m* and *n* are the number of half-waves in the  $x-$ and $y-$
directions, and $f_{m,n}$ is the natural frequency corresponding to the $(m,n)$
mode of vibration.

The corresponding relationships for a rectangular plate clamped on all
four edges are [1, 7]:

$$k_x = \frac{(2m+1)\pi}{l_x} \text{ for } m = 1,2,3\ldots\ldots\ldots \tag{2.68}$$

$$k_y = \frac{(2n+1)\pi}{l_y} \text{ for } n = 1,2,3\ldots\ldots\ldots \tag{2.69}$$

$$f_{m,n} = \left(1.8c_{Lp}h\right)\left\{\left(\frac{2m+1}{2l_x}\right)^2 + \left(\frac{2n+1}{2l_y}\right)^2\right\}, Hz \tag{2.70}$$

Equations (2.62) – (2.70) indicate that the quasi-longitudinal wave
speed along a plate is slightly higher than the longitudinal wave speed
along a slender rod. Besides, natural frequency $f_{m,n}$ of a plate clamped
along all four edges is considerably higher than that of a plate simply
supported along all the edges.

In practice, vibrating surface of a machine cannot be idealized as
rectangular thin plate with boundary conditions of clamping or simple
support. Then, one resorts to numerical analysis on a digital computer
making use of FEM and/or measurements, where use is made of the

orthogonality of normal modes. This is particularly true when one tries to evaluate sound power radiated by the surfaces of a typical machine like engine, gear box, automobile, etc., excited by multifarious time-variant forces.

## 2.7 Vibration Measurement

For harmonic time dependence $(e^{j\omega t})$, displacement $\xi$, velocity $u$ and acceleration $a$ are related as follows:

$$u = j\omega\xi \quad \text{and} \quad a = j\omega u = -\omega^2\xi \tag{2.71}$$

Thus, amplitudes of $u$ and $a$ would be $\omega$ and $\omega^2$ times that of the corresponding displacement. For a constant velocity spectrum, the displacement would decrease linearly with frequency and acceleration would increase linearly with frequency, as shown in Fig. 2.13.

Therefore, in general, displacement transducer is used for low frequency measurements and acceleration transducer (accelerometer) is preferred for high frequency measurements. For mid-frequency measurements and in general for wide-frequency spectrum measurements, velocity pickup may be used. However, accelerometer is the most commonly used vibration transducer; it has the best all-round characteristics.

Often, vibration, like sound, is measured in logarithmic units, decibels (dB), as follows.

$$\text{Displacement level, } L_d = 20 \ \log\left(\frac{\xi_{rms}}{\xi_{ref}}\right), \quad dB \tag{2.72}$$

$$\text{Velocity level, } L_u = 20 \ \log\left(\frac{u_{rms}}{u_{ref}}\right), \quad dB \tag{2.73}$$

$$\text{Acceleration level, } L_a = 20 \ \log\left(\frac{a_{rms}}{a_{ref}}\right), \quad dB \tag{2.74}$$

Here, as per international standards, the reference values are [1]:

$$\xi_{ref} = 10^{-12} m = 1 \ picometer \tag{2.75}$$

$$u_{ref} = 10^{-9}\, m\,/\,s = 1 \text{ nanometer/s} \tag{2.76}$$

$$a_{ref} = 10^{-6}\, m\,/\,s^2 = 1 \text{ micrometer/s}^2 \tag{2.77}$$

It may be noted that at $\omega = 1000$ rad/s, or at $f = 1000/2\pi$ $= 159.155$ Hz, all three levels will be equal. This determines the frequency at which all three lines intersect in Fig. 2.13.

The reference values indicated in Eqs. (2.75) – (2.77) are not always adhered to in the existing literature. Therefore, it is advisable to indicate the reference value used in prescribing vibration levels; for example, velocity level = 75 dB *re* $10^{-9}$ m / s.

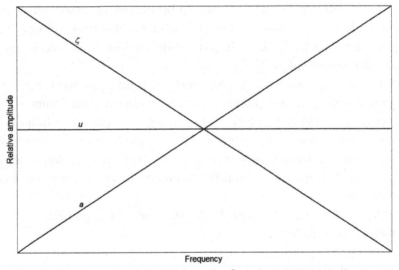

Fig. 2.13 Relative amplitudes of displacement ($\xi$), velocity (u) and acceleration (a)
(for constant velocity) .

Vibration levels may be added or subtracted in the same way as sound pressure levels, as illustrated before in Chapter 1.

**Example 2.5** Vibration velocity levels were measured to be 100, 90, 80 and 70 dB in the Octave bands centered at 31.5, 63, 125, and 250 Hz, respectively. Evaluate the total rms values of displacement, velocity and acceleration.

## Solution

We make use of Eqs. (2.72), (2.73), and (2.74) that relate the rms values of displacement, velocity and acceleration to the respective levels in dB, and Eq. (2.71) to evaluate displacement and acceleration from velocity.

Thus,

$$u_{rms} = 10^{-9} . 10^{L_u/20}, \quad \xi_{rms} = \frac{u_{rms}}{\omega}, \quad a_{rms} = \omega . u_{rms}, \quad \omega = 2\pi f$$

| Octave band Centre frequency f (Hz) | 31.5 | 63 | 125 | 250 | Total |
|---|---|---|---|---|---|
| Velocity level $L_u$(dB) | 100 | 90 | 80 | 70 | 100.5 |
| $u_{rms}$ (m/s) | $1.0 \times 10^{-4}$ | $3.16 \times 10^{-5}$ | $1.0 \times 10^{-5}$ | $3.16 \times 10^{-6}$ | $1.06 \times 10^{-4}$ |
| $\omega = 2\pi f$ (rad / s) | 196 | 393 | 785 | 1571 | - |
| $\xi_{rms} = u_{rms}/\omega$ (m) | $5.1 \times 10^{-7}$ | $8.04 \times 10^{-8}$ | $1.27 \times 10^{-8}$ | $2.01 \times 10^{-9}$ | $5.16 \times 10^{-7}$ |
| $a_{rms} = \omega . u_{rms}$ (m/s²) | 0.0196 | 0.0124 | 0.0078 | 0.0050 | 0.025 |

In the foregoing table, the total velocity level is determined by means of Eq. (1.37), and the total root mean square value of velocity is evaluated as follows:

$$u_{rms}(total) = \left[ \sum_{i=1}^{4} u_{rms}^2 (f_i) \right]^{1/2}$$

Similar expressions have been used for $\xi_{rms}(total)$ *and* $a_{rms}(total)$.

Incidentally, it may be noted that total values in the last column are practically equal to those in the first column where the velocity level is 100 dB; i.e., higher than those in other octave bands by 10 dB or more. This observation may be used effectively for noise and vibration control as will be demonstrated in the subsequent chapters.

Depending on the mode of transduction, different types of vibration transducers are available in the market; namely, eddy current displacement probes, moving element velocity pickups, piezoelectric accelerometers, etc.

Eddy current displacement probes are non-contacting displacement transducers with no moving parts and work right down to zero frequency. However, their higher frequency limit is about 400 Hz because, as indicated above, displacement often decreases with frequency. Being of the non-contacting type, the eddy-current probes are ideally suited for rotating machinery. However, their dynamic range is limited to 100:1 or 40 dB $(20 \log(100/1) = 40\ dB)$.

This limitation of dynamic range is also typical of the moving element velocity pick-ups. As these have to be in contact with the vibrating surface, their mass may alter the vibration that is supposed to be measured, particularly for thin sheet metal components. The lower frequency limit of these pick-ups is about 10 Hz because they operate above their mounted resonance frequencies. The moving elements are prone to wear and therefore the durability of these pickups is rather limited. Sensitivity to orientation and magnetic fields are additional concerns with the moving element velocity pick-ups.

Accelerometers have a very wide dynamic range $(> 120\ dB)$ as well as frequency range, although they fail as frequency tends to zero. Accelerometers are rugged transducers. Mostly they are of the piezoelectric type where sensing element is a polarized piezoelectric crystal or ferroelectric ceramic element. An electric charge is produced when this element is stressed in shear or in tension/compression. Fig. 2.14 shows a schematic of a compression-type piezoelectric accelerometer.

Acceleration of the mass element in Fig. 2.14 provides the inertial compressive force on the piezoelectric crystal, with the preloading spring ensuring that this force remains compressive throughout the oscillation. The compression type accelerometer is generally used for measuring high shock levels, whereas the shear type accelerometer serves a general purpose. Like velocity pick-up, accelerometer has to be in contact with the measuring surface and therefore has to be designed to be light and small enough not to interface with vibration that it is supposed to measure.

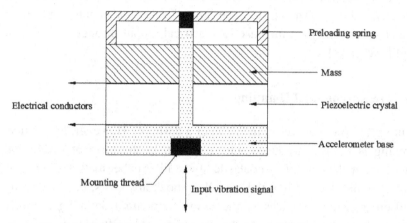

Fig. 2.14 Schematic of a compression-type piezoelectric accelerometer.

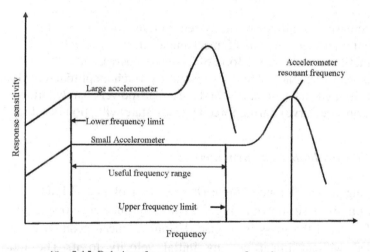

Fig. 2.15 Relatives frequency response of accelerometers.

Accelerometer's useful frequency range is limited on the lower frequency side by decreased sensitivity and on the higher frequency side by the resonance frequency of the accelerometer, as shown in Fig. 2.15. This resonance frequency is the result of the preloading spring interacting with the inertial mass. In general, larger accelerometers have larger sensitivity, but have a restricted useful frequency range. They may also interfere with the vibration surface through inertial loading as indicated above. Therefore, one must select a small and light accelerometer and then increase its electrical output through appropriate amplification [1].

## 2.8   Measurement of Damping

Figure 2.1 shows damper in the form of a dashpot. However, in practice, damping is rarely introduced into a system in the form of a dashpot. Often, one makes use of viscoelastic layers like rubber mats and wedges, fluid-dynamic devices like the so-called shock absorbers, etc. Damping coefficient of such devices or the in-built structural damping is rarely known or given; it needs to be measured [5]. Often it is inferred indirectly from different manifestations of damping; viz.,

(a) decreasing amplitude of the system in free vibration (see Fig. 2.2),
(b) continuous conversion of mechanical energy (potential as well as kinetic) into heat in the form of hysterisis losses, and
(c) resonance characteristics (the extent to which amplitude of vibration is limited at resonance and the half-power bandwidth of the resonance curve) during forced (steady state) vibration (see Fig. 2.3).

### 2.8.1   *Logarithmic decrement method*

Referring to the damped single-DOF system of Fig. 2.1 and the free vibration response shown in Fig. 2.2, the amplitude envelope is given by $x_0 e^{-\zeta \omega_n t}$ when the mass is displaced from its mean position by $x_0$ and released without imparting any initial velocity to it. The ratio of amplitudes for two successive peaks would then be

$$\frac{X_i}{X_{i+1}} = \frac{x_0 e^{-\zeta \omega_n t}}{x_0 e^{-\zeta \omega_n (t+T)}} = e^{\zeta \omega_n T} = e^{-\zeta \omega_n 2\pi / \omega_d}$$

$$= e^{2\pi \zeta / \left(1 - \zeta^2\right)^{1/2}} \simeq e^{2\pi \zeta} \tag{2.78}$$

because the damping ratio $\zeta$ is often much less than unity.

A popular measure of damping is Logarithmic Decrement, $\delta$, defined as natural logarithm of the ratio of successive amplitudes. Thus,

$$\delta = ln\left(\frac{X_i}{X_{i+1}}\right) \simeq 2\pi\zeta \tag{2.79}$$

Obviously, $\delta$ is easy to measure. Then, Eq. (2.79) can be used to get damping ratio, $\zeta$. Finally, damping coefficient c may be determined from Eqs. (2.2); i.e.,

$$c = c_c\zeta = 2m\omega_n\zeta \tag{2.80}$$

### 2.8.2 Half-power bandwidth method

It may be noted from the steady-state response curves of Fig. 2.3 that with increasing damping, the resonance amplitude decreases and the curve becomes more flat. This flat-ness, in terms of half-power bandwidth, may be used to evaluate the damping ratio $\zeta$ and thence the damping coefficient c. In fact, this method is a standard method for evaluation of the inherent or structural damping of materials (beams and plates) in flexural mode of vibration in terms of the loss factor $\eta$, which is a measure of to what extent strain fluctuations would lag behind the corresponding stress fluctuations. It is defined in terms of the complex character of the Young's modulus:

$$E = E_r + j\,E_i = E_r\left(1 + j\eta\right),\ \eta \equiv E_i\,/\,E_r \tag{2.81}$$

where $E_r$ is called the Storage Modulus,
  $E_i$ is called the Loss Modulus, and
  $\eta$ is called the Loss Factor.

For rubber-like viscoelastics, $E_r$, $E_i$ and $\eta$ are functions of frequency. These are evaluated by the standard Oberst's beam method [8].

When the system (or a cantilever beam) is excited by means of a wide-band random excitation, the typical response (vibration amplitude) consists of several peaks, one of which is shown in Fig. 2.16.

If $f_0$ is a resonance frequency with peak amplitude of $X_0$ and $f_1$ and $f_2$ are the frequencies on either side of $f_0$, as shown in Fig. 2.16, where

$X_1 = X_2 = 0.707 \, X_0$  (the power is proportional to vibration amplitude squared), then,

Normalized half-power bandwidth, $bw \equiv \dfrac{f_2 - f_1}{f_0}$          (2.82)

It may be shown that for small values of damping $\left(\zeta^2 \ll 1\right)$, different measures of damping are interrelated as follows [9]:

$$\zeta = \frac{\eta}{2} = \frac{bw}{2} = \frac{\delta}{2\pi} = \frac{1}{2Q}$$          (2.83)

Here Q is called the quality factor (in Electrical Filter theory), representing the non-dimensional amplitude or dynamic magnification factor $X/\left(F_0/k\right)$ at resonance $\omega = \omega_0 \left(\text{or}, \ f = f_0\right)$.

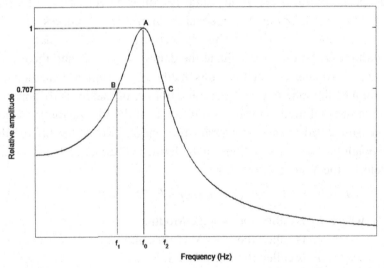

Fig. 2.16 Illustration of half-power bandwidth (points B and C are half-power points).

**Example 2.6** In an Oberst Beam test, the normalized half-power bandwidth of the first resonance peak is measured to be 0.16 for a particular alloy steel. If the same cantilever beam were disturbed and left to vibrate freely, what would be the ratio of two successive amplitudes of the beam in free vibration?

## Solution

The normalized half-power bandwidth $bw$ is given to be 0.16. Making use of Eqs. (2.83), the logarithmic decrement

$$\delta = \pi \, x \, bw = \pi \, x \, 0.16 = 0.5$$

Now, by definition of logarithmic decrement, the ratio of two successive amplitudes, $X_i / X_{i+1}$ is given by Eq. (2.79):

$$\frac{X_i}{X_{i+1}} = e^{\delta} = e^{0.5} = 1.65$$

Incidentally, making use of Eqs. (2.83) again, the loss factor $\eta$ of the alloy steel is given by

$$\eta = bw = 0.16$$

By comparison, loss factor of normal steel is of the order of 0.001. Thus, the loss factor of this alloy steel is 160 times that of ordinary steel. Therefore, the resonance amplitude of a plate made out of this alloy steel will be only 1/160 times that of normal steel plate. In other words, the alloy steel plate will be quieter by

$$20 \, \log 160 = 44.1 \quad dB$$

in free or resonant vibration, provided it has the same storage modulus $E_r$.

This is why sheet metal components of engines, compressors, etc. should be made out of materials with good inherent structural damping (high loss factor $\eta$), or better still, high loss modulus $E_i = \eta \, E_r$ (see Eq. 2.81).

## References

1. Norton, M. P., Fundamentals of Noise and Vibration, Cambridge University Press, Cambridge, (1989).
2. Yang, B., Theory of Vibration – Fundamentals, in Encyclopedia of Vibration (Ed. S. G. Braun) Academic Press, San Diego, pp. 1290-1299, (2002).

3. Lieven, N. A. J., Forced Response, in Encyclopedia of Vibration (Ed. S. G. Braun), Academic Press, San Diego, pp. 578-582, (2002).
4. Irwin, J. D. and Graf, E. R., Industrial Noise and Vibration Control, Prentice Hall, Englewood Cliffs, (1979).
5. Blevins, R. D., Formulas for Natural Frequency and Mode Shape, Van Nostrand Reinhold Co., New York, (1979).
6. Olson, M. F., Dynamical Analogies, Second Edition, Van Nostrand, Princeton, (1958).
7. Norton, M. P. and Drew, S. J., Radiation by Flexural Elements, in Encyclopedia of Vibration (Ed., S. G. Braun), Academic Press, San Diego, pp. 1456-1480, (2002).
8. Anon., Standard test method for measuring vibration-damping properties of materials, ASTM International Standard, E 756-04, (2004).
9. Ewins, D. G., Damping Measurement, in Encyclopedia of Vibration (Ed. S. G. Braun), Academic Press, San Diego, pp. 332-335, (2002).

## Problems in Chapter 2

**Problem 2.1** If the machine of the single-DOF system of Example 2.1 (m = 500 kg, k = 50968 N/m) were excited by a vertical unbalanced force of 100 N at a rotational speed of 1500 RPM (revolution for minute), then

(a) what would be the amplitude of vibration of the machine, and
(b) what would be amplitude of the force transmitted to the (rigid) foundation?

**[Ans.: (a) 8.14 micron, (b) 0.4153 $N$]**

**Problem 2.2** A single–DOF spring-mass-damper system shown in Fig. 2.1 has the following parameters: mass $m = 6$ *kg,* spring constant, $k = 15000$ *N/m.* Determine the damping coefficient, c of the dashpot from the observation that in free vibration, amplitude of oscillation of the mass decreases to 20 % of its displacement in seven consecutive cycles.

**[Ans.: 21.96 *Ns/m*]**

**Problem 2.3** If the system of Fig. 2.1 is executing free vibrations and its displacement $x(t)$ is given by the expression

$$x(t) = 0.002 \ e^{-6t} \left( \sin \ 10.4t + 1.732 \ \cos \ 10.4t \right),$$

determine the undamped natural frequency $\omega_n$ and damping ratio $\zeta$ of the system.

[**Ans.: 12.0** *rad/s* **and 0.5**]

**Problem 2.4**   For the system of Fig. 2.1, let $m = 100 \ kg$, $k = 100 \ N \ / \ mm$, $c = 10 \ Ns \ / \ cm$. Evaluate its damping ratio, undamped natural frequency, and amplitude at resonance for the force amplitude of 100 N.

[**Ans.: 0.158, 31.6** *rad/s* **and 3.16** *mm*]

**Problem 2.5**   The root-mean-square vibration acceleration values measured in different octave bands are listed below:

| Octave band centre frequency (Hz) | 31.5 | 63 | 125 | 250 | 500 | 1000 |
|---|---|---|---|---|---|---|
| RMS vibration acceleration (m/s$^2$) | 5.0 | 4.0 | 3.0 | 2.0 | 1.0 | 0.5 |

Calculate   (a) overall RMS acceleration in m/s$^2$ and decibels
             (b) overall RMS velocity in mm/s and decibels
             (c) overall RMS displacement in $\mu m$ and decibels

[**Ans.: (a) 7.433** $mm/s^2$ **and 137.4** *dB*;
**(b)  27.54** $mm/s$ **and 148.8** *dB*; **(c) 130.4** $\mu m$ **and 162** *dB* ]

# Chapter 3

# Vibration Control

A vibration problem generally involves a source of vibration, a dynamical (or vibratory) system and the response. The system may be looked at as 'transmission path' as is done in the noise control practice (see Chapter 4). Control of vibration, therefore, consists in modifying the source and/or the system.

Typical sources of vibration are unbalanced forces and moments in a reciprocating and rotating machinery, turbulent flow in pipes (particularly, at the bends and intersections), standing waves and surges. Therefore, vibration control at the source involves balancing of reciprocating and rotating machinery (engines, turbines, compressors, blowers, etc.), smoothening of flow in pipes (avoiding separation of boundary layer), proper lubrication at joints, etc. Often, there is a coupling between vibration and noise; reducing one would result in reduction of the other. That is why the vibration control and noise control are considered together in this textbook.

The system consists of inertias, isolator springs and dampers. Control of vibration involves proper design of the system (or transmission path) so as to reduce the vibration of machinery as well as transmissibility over the entire range of speeds or forcing frequencies.

Permissible vibration level for each part of a system in a dynamic setting is decided by requirements of functionality or comfort. International Standards Organization (ISO) provides various regulations and standards which are usually stated in terms of amplitude, frequency and, sometimes, duration of test. Fig. 3.1 is representative of the acceptable vibration levels.

As indicated in Chapter 2, vibration control requirement occurs in two classes of vibration: (a) isolation of a sensitive instrument from the support or base motion, and (b) isolation of a support or foundation from the unbalanced forces generated within a machine. Base excitation occurs during vehicle motion over an undulating surface, satellite launch, and in the operation of disk drives. The force transmissibility problem relates to machine mounts, engine mounts, machine tool vibration, etc.

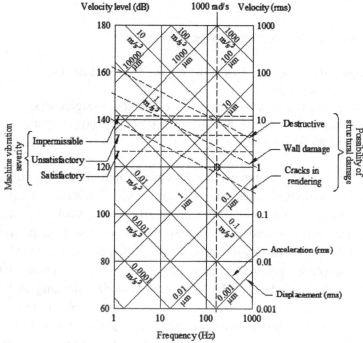

Fig. 3.1 Acceptable vibration levels. (Note that at 159.15 Hz or 1000 rad/s, displacement level, velocity level and acceleration level will all be equal to 120 dB, as indicated earlier in Fig. 2.13).

It is generally cost effective to control vibration at the source. Theory of vibration outlined in Chapter 2 suggests a number of vibration control measures. These are discussed at some length in the following sections.

## 3.1 Vibration Control at the Source

As indicated above, vibration can be controlled at the source by reducing the excitation. This may be done by

(a) reducing the rotational speed of the machine, if possible, without compromising the primary function of the machine;
(b) reducing the unbalance, which in turn may be effected by
    (i) precise machining of the rotor or crankshaft, and
    (ii) selecting a proper configuration (e.g., 6-cylinder inline engine is inherently balanced).

Similarly, the flow excitation may be reduced by

(a) reducing the flow speed by reducing the flow and/or increasing the area of cross section;
(b) smoothening the flow by means of tubular flow straighteners;
(c) streamlining the flow or avoiding the boundary layer separation by means of guide vanes.

Often, tall vertical chimneys or stacks are set into violent vibration by transverse oscillating forces caused by vortex shedding (Karman vortices) during strong wind. This may be effectively avoided by means of helical spoilers around the chimney that break the regular vortex pattern in the wake, thereby reducing the excitation dramatically. This is a good example of vibration control at the source. A variation of the same principle is made use of in order to minimize sloshing in large vertical, cylindrical liquid reservoirs. Perforated vertical plates are placed in the reservoir, parallel to, and away from, but near the container wall.

One common source of vibration (and noise) is dry friction caused by failure of lubrication. This causes high-frequency self excitation. This may be avoided or remedied by providing proper grease cups and keeping them under regular inspection.

A rotating shaft with a keyway causes parametric excitation due to periodical variation in transverse flexural stiffness. This may be remedied by providing two identical keyways on either side of the original keyway at $120°$ azimuthal locations. A precision job would completely nullify the parametric excitation that could have caused a whirling motion of the rotor, as explained in Chapter 2.

In general, the unbalanced forces and moments of a reciprocating engine crankshaft and turbine or compressor rotor cannot be reduced to zero. Therefore, balancing has to be done at site and/or at the manufacturer's end. Several techniques and machine have been developed over the year for balancing of rigid rotors. These include [1]

(a) pivoted-carriage balancing machine
(b) Gisholt-type balancing machine

A very large rotor cannot be mounted on any balancing machine. However, portable sets for carrying out the field balancing of both single and two-plane rotors are commercially available.

A rotor operating at a speed higher than its first critical speed undergoes a significant transverse deflection at this speed, and therefore is termed as flexible rotor. In a rigid rotor, the balancing masses are attached to neutralize the unbalanced forces and moments. However, in a flexible rotor, the balancing masses are designed to suitably modify the dynamic deflection characteristics of the rotor. This is effected by means of the modal balancing technique [2].

## 3.2 Vibration Isolators

Theory of vibration of single degree-of-freedom systems has been discussed in Chapter 2. Salient conclusions of the same regarding vibration isolation were as follows.

(i) For an isolator to yield low transmissibility, the natural frequency of the system should be much less than the forcing frequency, which is often equal to the rotational speed of the machine in corresponding units.

(ii) A sufficiently low natural frequency can be achieved by means of soft springs and/or inertia block.

(iii) Use of inertia block has the additional advantage of limiting the vibration levels of the machine.

(iv) Vibration of the machine and transmissibility at and around the resonance frequency may be controlled by means of damping. Excessive damping could, however, increase transmissibility at higher frequencies. Viscoelastic elements provide compliance as well as damping.

### 3.2.1  *Bonded rubber springs*

Natural frequency is inversely proportional to the square root of static deflection. The requirement of sufficiently low natural frequency results in large static deflection. Fig 3.2 indicates the range of static deflection for different types of isolators.

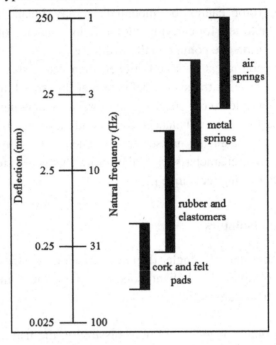

Fig. 3.2  Range of static deflection for different types of isolators.

Large static deflection results in mechanical instability. This may be avoided by means of bonded rubber springs. Most anti-vibration mounts in the market make use of these springs. As its name suggests, a bonded rubber spring is constructed by bonding the rubber to metal parts. In fact, one can make use of several of them in series, as shown schematically in Fig. 3.3. It may be noted that four identical rubber pads have been bonded to five thin metal plates. Stiffness of such a composite spring in compression will be roughly one-fourth of that of a single rubber pad, *EA / h*, where A is area of cross-section of the spring. The same would apply to the spring stiffness in shear. Thus, for the composite rubber spring of Fig. 3.3,

$$\text{Compressive or axial stiffness, } k_a \approx \frac{EA}{4h} \qquad (3.1)$$

$$\text{Torsional or shear stiffness, } k_s \approx \frac{GA}{4h} \qquad (3.2)$$

Here, h is thickness of each of the four rubber pads. It may be noted that thickness of thin metal plates does not influence the axial as well as shear stiffness of the composite bonded rubber spring shown in Fig. 3.3 because stiffness of the metallic sheet is higher than that of the rubber pad by several orders.

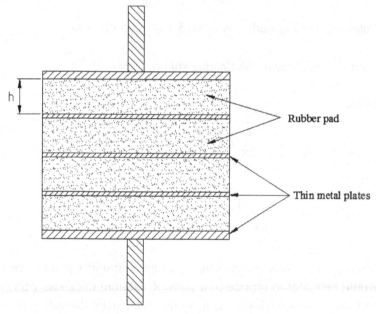

h

Rubber pad

Thin metal plates

Fig. 3.3 A composite bonded rubber spring.

**Example 3.1** A studio of total mass 40.5 tons is to be supported on nine composite bonded rubber springs of the type shown in Fig. 3.3 in order to ensure a floating floor with natural frequency of 8 Hz. This is required to avoid flanking transmission of structure-borne sound at 32 Hz upwards. If the surface area of each of the four rubber pads in each of the nine springs is $0.3 \text{ m} \times 0.3 \text{ m}$, and elastic modulus of rubber is $10^8 \text{ N/m}^2$, then evaluate thickness of each of the rubber pads. Assuming loss factor of rubber to be 0.2, evaluate transmissibility.

**Solution**

Stiffness of each spring

$$= \frac{EA}{4h} = \frac{10^8 \times (0.3 \times 0.3)}{4h} = \frac{0.0225 \times 10^8}{h} \text{ N/m}^2$$

Total stiffness of all nine springs,

$$k = \frac{9 \times 0.0225 \times 10^8}{h} = \frac{0.2025 \times 10^8}{h} \text{ N/m}^2$$

Total mass of the studio, $m = 40.5 \ T = 4.05 \times 10^4 \text{ kg}$

Natural frequency of the floating studio, $f_n = \frac{1}{2\pi} \left( \frac{k}{m} \right)^{1/2}$

Thus,

$$\frac{1}{2\pi} \left( \frac{0.2025 \times 10^8}{h \times 4.05 \times 10^4} \right)^{1/2} = 8$$

which gives

pad thickness, $h = \dfrac{0.2025 \times 10^8}{64 \times 4\pi^2 \times 4.05 \times 10^4} = 0.198 \text{ m} = 198 \text{ mm}$

This is unusually large from the mechanical stability point of view. It would be better to replace four pads of 198 mm thickness with eight pads of 99 mm thickness each, so that total static deflection remains unchanged.

As per Eq. (2.83), damping ratio, $\zeta = \dfrac{\eta}{2} = \dfrac{0.2}{2} = 0.1$

Frequency ratio, $r = \dfrac{f}{f_n} = \dfrac{32}{8} = 4$

Finally, use of Eq. (2.33) gives

$$\text{Transmissibility } TR = \left[ \frac{1+\left(2\times0.1\times4\right)^2}{\left(1-4^2\right)^2+\left(2\times0.1\times4\right)^2} \right]^{1/2} = 0.085 = 8.5\%$$

Incidentally, without damping, the transmissibility would be:

$$TR \ (\textit{without damping}) \ = \left| \frac{1}{1-16} \right| = 0.0667 = 6.67\%$$

Bonded rubber springs are available in various forms. Each is configured for a particular type of loading. For combined compression and shear loading, one uses anti-vibration mount (AVM) shown in Fig. 3.4.

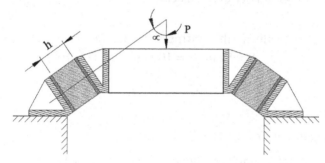

Fig. 3.4 Anti-vibration mount (AVM) for combined compression and shear loading [1].

Its stiffness for vertical loading is given by

$$k = 2A\left(G \sin^2\alpha + E_a \cos^2 \alpha\right)\!/h \tag{3.3}$$

Here A is the cross-sectional area of each rubber pad normal to the dimension h, and $E_a$ is the apparent Young's modulus in compression,

$$E_a = 6.12 \ G \tag{3.4}$$

Poisson's ratio, $v$, for rubber is nearly 0.5. Therefore, Young's modulus for rubber is given by

$$E = 2\left(1+v\right)G = 3G \tag{3.5}$$

Thus, the constraining effect of the end metal plates is to increase the apparent Young's modulus $E_a$ to more than double its basic value. The angle $\alpha$ in Fig. 3.4 is determined from the consideration of limiting the compressive strain to 0.2 and the shear strain to 0.36. If $x$ is the vertical deflection of the spring, then the accompanying shear deflection $x_s = x \sin \alpha$, and the compressive deflection $x_c = x \cos \alpha$. Therefore, the shear strain is $x \sin \alpha / h$ and the compressive strain is $x \cos \alpha / h$. In order to ensure that the two types of strain reach their limits simultaneously, $\alpha$ is fixed at arctan (0.36/0.2) or $60^0$ [1].

The anti-vibration mounts (AVM) of the type of Fig. 3.4 are used extensively under stationary installations like diesel generator (DG) sets, turbo-generators (TG), stationary compressors, etc. Often, several AVM's are used in parallel such that their centre of buoyancy coincides with the centre of gravity of the machine.

**Example 3.2** Evaluate the stiffness of the anti-vibration mount of Fig. 3.4 for $\alpha \approx 60^0$, h = 5 cm, A = 100 cm$^2$ and G = $10^6$ Pa.

**Solution**

Making use of Eqs. (3.4) and (3.3) we get

Apparent Young's modulus, $E_a = 6.12 \times 10^6$ Pa

Stiffness,

$$k = 2 \times \frac{100}{100 \times 100} \left( 10^6 \sin^2 \left( \pi/3 \right) + 6.12 \times 10^6 \cos^2 \left( \pi/3 \right) \right) \bigg/ \frac{5}{100}$$

$$= 2 \times 10^6 \left( \frac{3}{4} + 6.12 \left( \frac{1}{4} \right) \right) \bigg/ 5$$

$$= 0.912 \times 10^6 \text{ Pa.m}$$

$$= 912 \text{ kN/m}$$

$$= 912 \text{ N/mm}$$

### 3.2.2  *Effect of compliant foundation*

In the theory of vibration dealt with in Chapter 2, the support or foundation was implicitly assumed to be rigid. However, in practice, the foundation or support would invariably have a finite impedance. In other words, it would be compliant, not rigid.

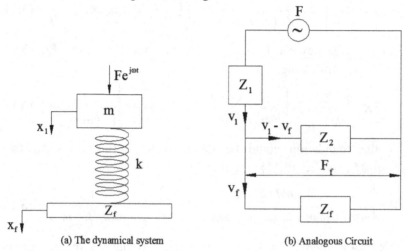

(a) The dynamical system       (b) Analogous Circuit

Fig. 3.5 Analysis of an isolator with compliant foundation.

Making use of the electro-mechanical analogies [3] discussed in chapter 2, for the dynamical system of Fig. 3.5 we have

$$Z_1 = j\omega m, \quad Z_2 = \frac{k}{j\omega}, \quad \left( v_1 = \dot{x}_1, \ v_f = \dot{x}_f \right) \tag{3.6}$$

$$Z_1 v_1 + Z_2 \left( v_1 - v_f \right) = F \tag{3.7}$$

$$Z_f v_f - Z_2 \left( v_1 - v_f \right) = 0 \tag{3.8}$$

Simultaneous solution of the linear algebraic equations (3.7) and (3.8) yields

$$v_f = \frac{Z_2 F}{Z_f \left( Z_1 + Z_2 \right) + Z_1 Z_2} \tag{3.9}$$

Therefore, the force transmitted to the foundation and transmissibility are given by

$$F_f = Z_f v_f = \frac{Z_f Z_2 F}{Z_f (Z_1 + Z_2) + Z_1 Z_2} \tag{3.10}$$

$$TR = \left| \frac{F_f}{F} \right| = \left| \frac{Z_f Z_2}{Z_f (Z_1 + Z_2) + Z_1 Z_2} \right|, \quad Z_1 = j\omega m, \ Z_2 = \frac{k}{j\omega} \tag{3.11}$$

As a consistency check, for a rigid foundation $(Z_f \to \infty)$, Eq. (3.11) reduces to the following:

$$TR = \left| \frac{Z_2}{Z_1 + Z_2} \right| = \left| \frac{k/j\omega}{j\omega m + k/j\omega} \right| = \left| \frac{1}{1 - m\omega^2/k} \right| = \left| \frac{1}{1 - (\omega/\omega_n)^2} \right| \tag{3.12}$$

If the foundation could be idealized as a free mass $M$, then $Z_f = j\omega M$, and Eq. (3.11) would yield

$$TR = \left| \frac{(j\omega M (k/j\omega))}{j\omega M (j\omega m + k/j\omega) + (j\omega m)(k/j\omega)} \right| = \left| \frac{1}{1 - m\omega^2/k + m/M} \right| \tag{3.13}$$

TR would tend to infinity at the natural frequency of the system, given by

$$1 - \frac{m\omega_n^2}{k} + \frac{m}{M} = 0$$

whence

$$\omega_n = \left( \frac{k(M+m)}{Mm} \right)^{1/2} = \left\{ \frac{k}{m} \left( 1 + \frac{m}{M} \right) \right\}^{1/2} \tag{3.14}$$

Thus, the natural frequency of a system with free inertial foundation would be higher than that with a rigid foundation.

For a compliant foundation, the real effectiveness of a vibration isolator (spring of stiffness $k$ in Fig. 3.5) is not represented by transmissibility. A more appropriate metric in this case would be the ratio of the force transmitted to the foundation with the spring and that without the spring $(Z_f \to \infty)$. Let us denote it by force ratio, FR. Thus,

$$FR = \left| \frac{F_f \text{ with isolator spring}}{F_f \text{ without isolator spring}} \right| = \left| \frac{\dfrac{Z_f Z_2 F}{Z_f(Z_1 + Z_2) + Z_1 Z_2}}{\dfrac{Z_f F}{Z_1 + Z_f}} \right| \qquad (3.15)$$

This simplifies to

$$FR = \left| \frac{Z_1 + Z_f}{Z_1 + Z_f (1 + Z_1/Z_2)} \right| \qquad (3.16)$$

It may be noted that for a rigid foundation $(Z_f \to \infty)$, Force Ratio FR reduces to Transmissibility TR; i.e., Eq. (3.12). In other words, only for rigid support or foundation, do the two metrics match.

**Example 3.3** The floating floor construction of Example 3.1 presumes a rigid foundation. If the foundation were compliant and could be modeled as a free mass of 200 tons, then what would be the transmissibility and the force ratio?

**Solution**

Transmissibility for a free mass foundation is given by Eq. (3.13), with

$$m = 4.05 \times 10^4 \text{ kg,}$$

$$k = \frac{0.2025 \times 10^8}{0.198} = 1.023 \times 10^8 \quad \text{N/m}$$

$$\omega = 2\pi \times 32 = 201.1 \quad \text{rad/s}$$

$$M = 200 \times 10^3 = 2.0 \times 10^5 \quad \text{kg}$$

Thus, neglecting damping, transmissibility is given by

$$TR = \left| \frac{1}{1 - 16.0 + 0.2} \right| = 0.0676 = 6.76\%$$

This may be noted to be only marginally higher than that for rigid foundation (6.67%). So, compliance of the foundation is of no significance in this case.

Now, Force Ratio, FR, is given by Eq. (3.16), where

$$Z_1 = j\omega m, \quad Z_f = j\omega M \quad and \quad Z_2 = k/j\omega$$

Thus, $\quad FR = \left| \dfrac{m+M}{m+M\left(1-\omega^2 m/k\right)} \right| = \left| \dfrac{4.05+20}{4.05+20(1-16)} \right| = 0.081$

This represents the real effectiveness of the set of isolator springs.

### 3.2.3 *Pneumatic suspension*

It has been shown before that for low transmissibility the natural frequency of the system should be much smaller than the forcing frequency. However, use of springs would result in large static deflection and mechanical instability. Therefore, low frequency isolation is not feasible with metallic springs or viscoelastic pads. Fortunately, pneumatic suspension does not suffer from this weakness; it can yield good isolation with little static deflection even if the forcing frequency tended to zero!

Making use of the adiabatic gas law, it can be shown that stiffness of an air spring (a piston with a surge tank (see Fig. 3.6 for a basic configuration)), stiffness is given by [1]

$$\text{Single acting:} \quad k \approx \gamma A^2 \, p_0 / V_0 \tag{3.17}$$

$$\text{Double acting:} \quad k \approx \gamma A^2 \left( \frac{p_1}{V_1} + \frac{p_2}{V_2} \right) \tag{3.18}$$

where A is the load bearing area, $p_0$ is the equilibrium pressure and $V_0$ is volume of the air in the cylinder and surge tank of a single acting air spring shown in Fig. 3.6(a); $p_1, p_2$ and $V_1, V_2$ are the corresponding pressures and volumes of double acting spring shown in Fig. 3.6 (b).

Force transmissibility or motion transmissibility (see Fig. 3.7) for rigid foundation can be obtained by means of Eq. (3.12) making use of the air spring stiffness from Eq. (3.18).

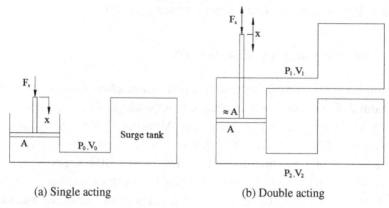

(a) Single acting                    (b) Double acting

Fig. 3.6  Schematic of pneumatic suspension (adopted with permission from Mallik [1]).

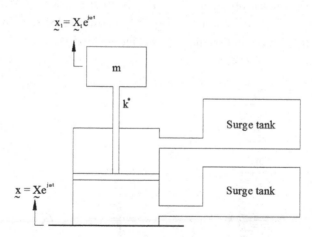

Fig. 3.7  Motion transmissibility of a double-acting air spring (adopted with permission from Mallik [1]).

Damping can be introduced into the pneumatic suspension system through capillary passages connecting the surge tanks with the cylinder. For analysis of such a system, the reader is referred to Mallik [1].

The stiffness expressions (3.17) and (3.18) hold the key for designer who can select appropriate values of $A$, $p_0$ and $V_0$ to obtain the required (low enough) stiffness and adequate load bearing capacity $(Ap_0)$. Obviously, use of a large surge tank mould ensure low stiffness without compromising the load bearing capacity. But then, there would invariably be a practical constraint on space availability.

### 3.3    Dynamic Vibration Absorber (DVA)

A spring-mass (and damper) pair attached to a machine excited at or near its natural frequency as a mechanical appendage so as to absorb its energy is called a Dynamic Vibration Absorber (DVA). Schematic of such an absorber is shown in Fig. 3.8(a). It was invented by Frahm at the beginning of the twentieth century, i.e., about 100 years ago. Its principle may be better understood by making use of analogous circuit [3] shown in Fig. 3.8 (b). Here $m$ is the mass of the vibrating machine (element 2), $k$ is the stiffness of the existing isolator spring (element 1), $m_a$ is the DVA mass (element 4) and $k_a$ is stiffness of the DVA spring (element 3). For convenience, Fig. 3.8 shows velocities instead of displacements. To recall the analogous relationships [4],

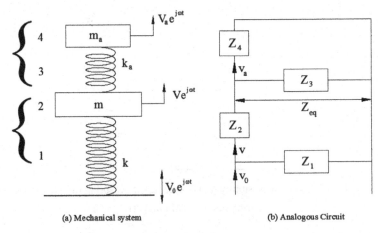

(a) Mechanical system                    (b) Analogous Circuit

Fig. 3.8 Illustration of the function of a dynamic vibration absorber (DVA).

$$v = j\omega x, \quad Z_1 = k/j\omega, \quad Z_2 = j\omega m, \quad Z_3 = k_a/j\omega, \quad Z_4 = j\omega m_a \qquad (3.19)$$

Subscript 'a' denotes absorber, or dynamic vibration absorber (DVA).

In Fig. 3.8 (b), $Z_3$ and $Z_4$ are in parallel. Therefore $Z_{eq}$, the equivalent impedance of the DVA on the machine mass $m$ is given by

$$Z_{eq} = \frac{Z_3 Z_4}{Z_3 + Z_4} = \frac{\dfrac{k_a}{j\omega} \cdot j\omega m_a}{\dfrac{k_a}{j\omega} + j\omega m_a} = \frac{j\omega m_a}{1 - \omega^2 m_a / k_a} \qquad (3.20)$$

Clearly, if $k_a/m_a = \omega^2$, i.e., if the natural frequency of the absorber is equal to the forcing frequency, then $Z_{eq} \to \infty$. This would open up the lower loop in Fig. 3.8 (b), making $v$, velocity of the machine mass $m$, tend to zero. This is the principle underling the dynamic vibration absorber.

At the tuned frequency, $\omega = \left(k_a/m_a\right)^{1/2}$, the machine (mass m) would not move $(v=0)$. It would be subjected to a dynamic force $Z_1 v_0$ from below (the support side) and an equal and opposite force $Z_3 v_a$ from above (see Fig. 3.8). Thus,

$$Z_3 v_a = Z_1 v_0 \qquad (3.21)$$

$$v_a = \frac{Z_1}{Z_3} v_0 = \frac{k}{k_a} v_0 = \frac{k}{m_a \omega^2} v_0 \qquad (3.22)$$

Thus, DVA mass $m_a$ would have considerable motion (velocity $v_a$). In order to keep the DVA motion within limits, its mass $m_a$ should not be too small.

A tuned vibration absorber $\left(\omega = \left(k_a/m_a\right)^{1/2}\right)$ can also be used to reduce transmissibility of the original system, $|v/v_0|$, at frequency ratio, $r \left(\equiv \omega/\omega_n, \ \omega_n = \left(k/m\right)^{1/2}\right)$ for $r > 1$. For this purpose, it can be proved that the effective range of a dynamic vibration absorber around its tuned frequency is proportional to the mass of the absorber and is governed by the approximate expression [3]:

$$\frac{\omega_2 - \omega_1}{\omega_a} = \frac{r_2 - r_1}{r_a} \simeq \frac{m_a}{m} \frac{1}{e}, \ r_a = \frac{\omega_a}{\omega_n} = \frac{\left(k_a/m_a\right)^{1/2}}{\left(k/m\right)^{1/2}} \qquad (3.23)$$

where e is an arbitrary large number representing the designed isolation. Thus, if one wishes to have a transmissibility equal to one-fifth that of the corresponding straight-through system (without DVA) over a 5% variation in the forcing frequency $r_a$ of 6, then

$$e = \frac{1}{1/5} = 5$$

$$\frac{m_a}{m} \simeq \frac{r_2 - r_1}{r_a}.e = \frac{5}{100} \times 5 = 0.25$$

$$\frac{k_a/m_a}{k/m} = (6)^2 = 36$$

Thus, equations (3.23) adjust $m_a$ and $k_a$ in terms of the given $m$ and $k$, completing thereby the design of the DVA.

Often, a DVA is designed to control the resonant vibration of a given machine or rotor or a sub-assembly. Then, $\omega = \omega_n = (k/m)^{1/2}$, and Eq. (3.22) reduces to

$$v_a = \frac{m}{m_a} \, v_0 \tag{3.24}$$

Obviously, $m_a$ must be large enough to limit its motion; otherwise, we would end up transferring the vibration problem from the machine to the DVA mass!

Figure 3.8 illustrates the use of a DVA for reducing the velocity transmissibility $(v/v_0)$ of a sensitive instrument of mass $m$. However, as shown in Ref. [3], the same principle as well as Eqs. (3.23) hold for reduction of the force transmissibility and amplitude of vibration of mass m acted upon by an oscillating force.

**Example 3.4** Design an undamped dynamic vibration absorber (evaluate $m_a$ and $k_a$) in order to reduce velocity transmissibility to one-fifth of the original value of 20% for a precision instrument weighing 10 kg mounted by means of a spring (see Fig. 3.8) on a surface vibrating vertically at about 100 Hz, with possible variation of 5%. Also, evaluate stiffness $k$ of the main spring

**Solution**

For the original (main) system, $TR = \left| \dfrac{1}{1 - r^2} \right| = 0.2$

whence,

$$r = 2.45$$

or

$$\omega\Big/\omega_n = f/f_n = 2.45 \Rightarrow f_n = \frac{f}{2.45} = \frac{100}{2.45} = 40.82 \text{ Hz}$$

Now,

$$\frac{1}{2\pi}\left(\frac{k}{m}\right)^{1/2} = f_n$$

or

$$\frac{1}{2\pi}\left(\frac{k}{10}\right)^{1/2} = 40.82 \Rightarrow k = \left(2\pi \times 40.82\right)^2 \times 10 = 6.58 \times 10^5 \quad \text{N/m}$$

Besides, Eq. (3.23) gives

$$\frac{\omega_2 - \omega_1}{\omega_a} = \frac{m_a}{m}\frac{1}{e}, \quad e = \frac{100}{20} = 5$$

Thus,

$$0.05 = \frac{m_a}{10}\cdot\frac{1}{5} \quad \Rightarrow \quad m_a = 2.5 \text{ kg}$$

A tuned dynamic vibration absorber would have its natural frequency equal to the forcing frequency. Thus,

$$\frac{1}{2\pi}\left(\frac{k_a}{m_a}\right)^{1/2} = 100 \Rightarrow k_a = \left(2\pi \times 100\right)^2 \times 2.5 = 9.87 \times 10^5 \quad \text{N/m}$$

Thus, a dynamic vibration absorber (DVA) with $m_a = 2.5$ kg and $k_a = 9.87 \times 10^5$ N/m would reduce the velocity transmissibility from 0.2 to $0.2 \times 0.2 = 0.04 = 4\%$. In other words, such a DVA would reduce amplitude of vibration of the instrument to 4% of that of the vibrating support.

One should use damping in parallel with spring, or combine the two elements into a viscoelastic element. As a matter of fact, most DVA's incorporate damping in one way or the other. Design of optimally damped DVA's may be found in several textbooks as well as research papers – see, for example, references [1, 5].

Figure 3.8 is just a schematic of a dynamic vibration absorber or vibration neutralizer or tuned mass damper. In practice, the primary vibrating system and the DVA may take a variety of shapes and forms. Some of the well-known applications are [1, 5]:

(a) the so-called stock bridge damper, widely used to reduce wind-induced vibration in the overhead power transmission lines;
(b) absorber for high-rise buildings, for suppressing primarily the contribution of the first vibration mode in wind-induced oscillations;
(c) pendulum-like DVA's applied to high television towers;
(d) devices used to
    (i) stabilize ship roll motion,
    (ii) attenuate vibrations transmitted from the main rotor to the cockpit of helicopters,
    (iii) improve machine tool operation conditions,
    (iv) reduce the dynamic forces transmitted to an aircraft due to high rates of fire imposed on the canon motion,
    (v) control torsional vibration of internal combustion engines and other rotating systems,
    (vi) attenuate ship roll motion (gyroscopic DVA);
(e) centrifugal pendulum vibration absorber.

This subject is not discussed further here because the scope of this text book is limited to control of vibrations that would lead to noise radiation.

## 3.4 Impedance Mismatch to Block Transmission of Vibration

It has often been observed that the unbalanced forces transmitted to the foundation or support structure travel long distance through the structure as structure-borne sound and radiate sound. This is also known as flanking transmission.

Therefore, reduction in noise radiation calls for

(a) control of unbalanced forces and moments at the source
(b) design of vibration isolators for minimal transmissibility
(c) blocking transmission of vibration in the support structure by means of
    (i) impedance mismatch (reflection of power flux)
    (ii) structural damping (absorption or dissipation of power flux)

### 3.4.1 *Viscoelastic interlayer*

Impedance mismatch is effected by sudden change in characteristic impedance along the transmission path. Common examples of the impedance mismatch are use of gaskets, washers, viscoelastic pads, etc., between two metallic layers or surfaces. The characteristic impedance Y (product of material mass density, $\rho$, and sound speed, $c$) of a viscoelastic material like rubber is several orders lower than that of a metal like steel.

It can be shown that the dynamic power transmission coefficient across an interface of two materials is given by [6]

$$\tau = \frac{W_t}{W_i} = \frac{4Y_1 Y_2}{\left(Y_1 + Y_2\right)^2}, \ Y_1 = \rho_1 c_1, \ Y_2 = \rho_2 c_2 \qquad (3.25)$$

where $W_t$ and $W_i$ represent the incident power flux and transmitted power flux, respectively (see Fig. 3.9).

In order to appreciate that $\tau$ is always less than unity, we may rewrite $\tau$ of Eq. (3.25) as

$$\tau = 1 - \left(\frac{Y_1 - Y_2}{Y_1 + Y_2}\right)^2 \qquad (3.26)$$

Thus,

$$\tau \ll 1 \ \ if \ \ Y_1 \ll Y_2 \ \ and \ also \ if \ \ Y_1 \gg Y_2 \qquad (3.27)$$

In other words, transmission of dynamic power across an interface of two materials with vastly different characteristic impedances is very poor. This illustrates the principle and use of impedance mismatch.

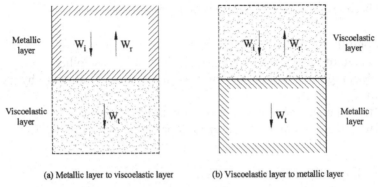

(a) Metallic layer to viscoelastic layer        (b) Viscoelastic layer to metallic layer

Fig. 3.9  Illustration of the principle of impedance mismatch.

Incidentally, symmetric nature of Eq. (3.26) shows that the power transmission coefficients for Figs. 3.9 (a) and (b) would be the same.

Making use of the conservation of energy (or power flux) at the interface of two surfaces we can see that

$$W_i - W_r = W_t \;\Rightarrow\; \frac{W_t}{W_i} = 1 - \frac{W_r}{W_i}, \;\; \frac{W_r}{W_i} = \left(\frac{Y_1 - Y_2}{Y_1 + Y_2}\right)^2 \qquad (3.28)$$

This shows that the effect of the impedance mismatch is to reflect a substantial part of the incident power back to the source. In other words, principle of impedance mismatch is the basis of reflective or reactive dynamical filters.

Figure 3.10 shows the use of a viscoelastic layer inbetween two metallic layers in order to block transmission of vibration (longitudinal waves, to be precise). This arrangement is very common in the mechanical engineering practice.  It represents a pair of impedance mismatch interfaces: A-A and B-B. However its overall transmission coefficient $\bar{\tau}$ is not as low as $\tau^2$, where $\tau$ is the transmission coefficient of each of the two individual interfaces A-A and B-B. In fact, it would be nearly unity except for thick and very compliant interlayers and at high frequencies. This is because the vibrational energy reflected at the interface B-B, comes back substantially after reflection from the interface A-A, and undergoes a series of back and forth reflections between the two interfaces before it gets absorbed or dissipated as heat within the viscoelastic layer. That is why, the intermediate layer must have viscosity (rather, structural damping) as well as elasticity.

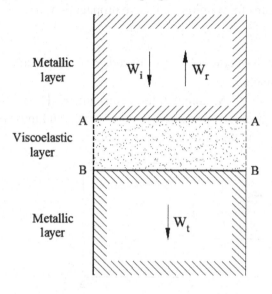

Fig. 3.10 Blockage of vibration transmission.

Unlike $\tau$ of an individual interface, $\bar{\tau}$ is a function of frequency as well as thickness of the viscoelastic interlayer. In general, its effectiveness would improve ($\bar{\tau}$ would decrease) with increased thickness and higher forcing frequency.

### 3.4.2 *Effect of blocking mass on longitudinal waves*

Impedance mismatch for blocking longitudinal waves in beams (rods) and plates can also be created by means of a blocking mass, which represents a couple of changes in cross-section, similar to the two interfaces in the case of a viscoelastic interlayer in the foregoing sub-section. It can be shown that the transmission coefficient of a blocking mass is given by [7]

$$\tau = \frac{1}{1 + \omega^2 M^2 / 4Y^2} \tag{3.29}$$

where $\omega$ is the forcing frequency of the longitudinal wave, $M$ is the blocking mass (see Fig. 3.11), and $Y$ is the characteristic impedance of the rod (or beam) for longitudinal wave propagation. It is given by [7]

$$Y = A'(\rho E')^{1/2} \tag{3.30}$$

where $A'$ = cross-sectional area $A$ for rods or beams, and thickness $h$ for plates,

$E'$ = Young's Modulus $E$ for beams and $E/(1-v^2)$ for plates,

$v$ and $\rho$ are, respectively, Poisson's ratio and mass density of the material of the beam or plate.

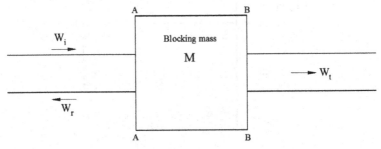

Fig. 3.11 Effect of blocking mass on vibration transmission.

A more common metric of the longitudinal wave transmission is Transmission Loss, TL. It is defined as

$$TL = 10 \ \log(W_i/W_t) = 10 \ \log(1/\tau) = -10 \ \log(\tau), \ dB \tag{3.31}$$

*TL* of a blocking mass shown in Fig. 3.11 may be obtained by combining Eqns. (3.29) and (3.31). Thus,

$$TL = 10 \ \log(1 + \omega^2 M^2/4Y^2) \simeq 20 \ \log(\omega M/2Y) = 20 \ \log(|Z|/2Y) \tag{3.32}$$

The approximate expression in Eq. (3.32) is for high frequencies and/or a large blocking mass such that $\omega M \gg 2Y$, and $Z = j\omega M$ is the inertive impedance of mass $M$.

Incidentally, the corresponding expression for the viscoelastic layer arrangement of Fig. 3.10 is given by

$$TL \simeq 20 \ \log\left|1 + \frac{Y}{2Z}\right| \simeq 10 \ \log(1 + \omega^2 Y^2/4k^2) \tag{3.33}$$

where $k$ is stiffness (inverse of compliance) of the viscoelastic layer; $k = EA/l$, $E$ is Elastic modulus, $A$ is area of cross-section, and $l$ is thickness of the viscoelastic layer. If we wish to consider loss factor of the layer as well, then $E$ and hence $Z$ would be complex.

**Example 3.5** A PVC gasket is used to isolate two steel surfaces. If the mass density and Young's modulus of the gasket are 1400 kg/m$^3$ and $2.4 \times 10^9$ N/m$^2$, respectively, and those of steel are 7800 kg/m$^3$ and $2.07 \times 10^{11}$ N/m$^2$, respectively, then evaluate

(a) characteristic impedance of the two materials;
(b) transmission coefficient and TL at each of the two interfaces for longitudinal waves; and
(c) TL due to the gasket of 3 cm thickness (see Fig. 3.10) at 10 Hz, 100 Hz and 1000 Hz.

**Solution**

Characteristic impedance, $Y = \rho c = \rho \left( \dfrac{E}{\rho} \right)^{1/2} = (E\rho)^{1/2}$

$Y$ of steel, $Y_1 = \left( 2.07 \times 10^{11} \times 7800 \right)^{1/2} = 4.02 \times 10^7 \ kg / \left( m^2 s \right)$

$Y$ of PVC, $Y_2 = \left( 2.4 \times 10^9 \times 1400 \right)^{1/2} = 1.83 \times 10^6 \ kg / \left( m^2 s \right)$

Use of Eq. (3.25) gives:

Transmission coefficient,

$$\tau = \frac{4 Y_1 Y_2}{\left( Y_1 + Y_2 \right)^2} = \frac{4 \times \left( 4.02 \times 10^7 \right) \left( 1.83 \times 10^6 \right)}{\left( 4.02 \times 10^7 + 1.83 \times 10^6 \right)^2}$$

$$= 0.166$$

Transmission loss, $TL = 10 \ \log \left( 1/0.166 \right)$

$$= 7.8 \ dB$$

Stiffness per unit area of the *3 cm* gasket, $k = E/l$

$$= \frac{2.4 \times 10^9}{3/100}$$

$$= 8.0 \times 10^{10} \ N/m^3$$

Making use of Eq. (3.33), *TL* due to the gasket in Fig. 3.10 is given by

$$TL = 10 \ \log\left\{1 + \frac{\omega^2 Y_1^2}{4k^2}\right\}, \quad \omega = 2\pi f$$

$$TL(f) = 10 \ \log\left\{1 + \left(\frac{2\pi f \times 4.02 \times 10^7}{2 \times 8.0 \times 10^{10}}\right)^2\right\}$$

$$= 10 \ \log\left\{1 + 2.49 \times 10^{-6} f^2\right\}$$

Thus, $TL = 0.0$ at 10 *Hz*, 0.1 *dB* at 100 *Hz*, and 5.4 *dB* at 1000 *Hz*. It may noted that TL of thin soft inter-layers is negligible at lower frequencies. In other words, soft inter-layers would act as good filters only at higher frequencies.

The same expression (3.32) would yield TL of a blocking rotor of polar mass moment of inertia J for torsional shear-wave transmission along a circular torsional member like shaft. Then,

$$Z = j\omega J \ and \ Y = I_p (\rho \ G)^{1/2}$$

where $I_p$ is cross-sectional polar moment of inertia of the shaft, and G is the shear modulus of the shaft material.

**Example 3.6** It is proposed to block transmission of torsional shear waves along a steel shaft of 50 mm diameters by means of a rotor of 500 mm diameter, 100 mm thick, made of steel. Estimate the resulting TL for torsional waves at 125 Hz.

## Solution

For steel,

Young's modulus, $E = 2.07 \times 10^{11}\ N/m^2$

density, $\rho = 7800\ kg/m^3$

Poisson's ratio, $v = 0.29$

Shear modulus is given by

$$G = \frac{E}{2(1+v)} = \frac{2.07 \times 10^{11}}{2(1+0.29)} = 0.8 \times 10^{11}\ N/m^2$$

Cross-sectional polar moment of inertia,

$$I_p = \frac{\pi}{32}\,d^4 = \frac{\pi}{32} \times (0.05)^4$$

$$= 6.136 \times 10^{-7}\ m^4$$

$$Y_t = I_p\,(\rho G)^{1/2} = 6.13 \times 10^{-7} \times \left(7800 \times 0.8 \times 10^{11}\right)^{1/2} = 15.313$$

Polar mass moment of inertia,

$$J = md^2 \Big/ 8 = \frac{\pi}{32} \times (0.5)^4 \times 7800 \times 0.1 = 4.786\ kg.m^2$$

$$\omega = 2\pi f = 2\pi \times 125 = 785.4\ rad/s$$

Finally,

the torsional wave $TL = 10\ \log\left[1 + \dfrac{\omega^2 J^2}{4 Y_t^2}\right]$

$$= 10\ \log\left[1 + \left(\frac{785.4 \times 4.786}{2 \times 15.313}\right)^2\right]$$

$$= 41.8\ dB$$

### 3.4.3  *Effect of blocking mass on flexural waves*

Noise radiation from vibrating beams and plates is due to flexural or transverse vibration. Here, the motion is perpendicular to the beam (or plate) while the wave (or disturbance) moves along the axis of the beam. The governing equations of flexural waves in beams and plates, and the concepts of flexural wave number, propagating (farfield) waves, evanescent (nearfield) waves, etc., were dealt with in Chapter 2 (see Section 2.6). Now, if there is a lumped blocking mass (line mass) $M$ located at $x = 0$ on an infinite uniform beam (rather, in the centre of two semi-infinite beams), then the farfield power transmission coefficient is given by [7]

$$\tau = \frac{\left(\bar{\mu}\bar{k} + 4\right)^2}{\left(\bar{\mu}\bar{k}\right)^2 + \left(\bar{\mu}\bar{k} + 4\right)^2} \tag{3.34}$$

where

$\bar{k} = \sqrt{12}\ k\,r$  is a nondimensional wave number,

$\bar{\mu} = \dfrac{M}{\sqrt{12}\rho A r}$  is a nondimensional blocking mass,

$r = \left(I/A\right)^{1/2}$  is the radius of gyration of the beam,

$k = \left(\dfrac{\rho A\,\omega^2}{EI}\right)^{1/4}$  is the flexural wave number, and

$\rho$, $A$ and $I$ are density, cross-sectional area and moment of inertia of the beam.

The corresponding relationships for plates with stiffeners may be found in references [8, 9]. Each stiffener is a lumped line mass. It not only provides additional stiffness against bending, but also reflects substantial portion of the flexural wave energy back to the source. Thus, stiffeners on hull plates of ships and submarines help in blocking of the flexural waves excited at the engine mounts or at the propeller thrust bearings, thereby reducing the noise emitted by the hull plate out into the sea or into the passenger cabins.

## 3.5 Damping Treatments for Plates

As indicated before, often the sheet metal components generate lot of noise. For example, the real sources of noise in an internal combustion engine is the sharp pressure fluctuations in the cylinder due to combustion. However, most of noise radiation occurs from the sheet metal components like oil pan, valve covers, gear covers, the intake and exhaust manifolds, etc. These components act as mechanical loudspeakers. This combustion–excited body noise may be reduced either by making these components out of materials with high loss modulus, or by means of damping treatments. Almost all practical damping materials are polymeric plastics or elastomers [10].

As noise is radiated primarily by flexural vibration, damping treatments or viscoelastic laminae are used for absorbing the energy of flexural waves in thin plates.

Damping treatments are of two types:

(a) free or unconstrained layer damping (see Fig. 3.12a)
(b) sandwich or constrained layer damping (see Fig. 3.12b)

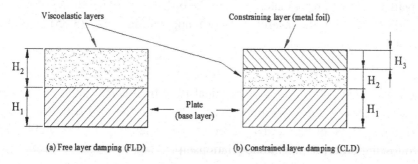

(a) Free layer damping (FLD)        (b) Constrained layer damping (CLD)

Fig. 3.12 Two types of damping treatments.

In the free layer damping (FLD), the flexural wave energy is dissipated through the mechanism of extensional deformation, whereas in the constrained layer damping (CLD), the wave energy is dissipated through shear deformation.

### 3.5.1 *Free layer damping (FLD) treatment*

The FLD treatment often consists of a homogeneous, uniform adhesive layer of mastic deadener. Its overall loss factor is given by [1, 11]

$$\eta \approx \frac{\eta_{E2}e_2 h_2 \left(3 + 6h_2 + 4h_2^2\right)}{1 + e_2 h_2 \left(3 + 6h_2 + 4h_2^2\right)} \tag{3.35}$$

where

$\eta_{E2}$ = loss factor of the viscoelastic layer in longitudinal deformation

$e_2 = E_2/E_1,\ h_2 = H_2/H_1,\ e_2 h_2 \ll 1$

$E_2$ = storage modulus of the viscoelastic layer

$E_1$ = Young's modulus of the base layer (plate)

$H_2$ = thickness of the viscoelastic layer

$H_1$ = thickness of the base layer (plate)

The numerator of Eq. (3.35) indicates that in order to increase the overall loss factor, we need to increase $\eta_{E2}\ E_2$ ; *i.e.*, the loss modulus of the viscoelastic layer, not its loss factor alone. Loss modulus of the best available commercial materials is of the order of $1\ GPa\left(=10^9\ N/m^2\right)$.

For most of the useful range of applications, Eq. (3.35) may be replaced with [11]

$$\eta \approx 14\ \eta_{E2}\ e_2\ h_2^2 \tag{3.36}$$

$\eta$ is less than $0.4\eta_E$ for most practical FLD treatments unless we make use of a very thick treatment, $H_2 \gg H_1$ in Fig. 3.12(a). Normally $h_2 = H_2/H_1$ is of the order of unity, and then Eq. (3.35) yields the following order-of-magnitude relationship:

$$\eta \sim \eta_{E2}\ e_2\ h_2^2 \sim 0.01\ \eta_E \tag{3.37}$$

Therefore, the FLD or unconstrained layer damping treatment leads to very little damping, and combined with its disadvantage of additional weight, it is rarely a cost effective solution.

**Example 3.7** A 2-mm thick aluminium plate is lined with a 2-mm thick viscoelastic layer of $E = 2 \times 10^9 (1 + j\ 0.5) N/m^2$. Calculate the composite loss factor of the lined plate.

**Solution**

We make use of Eq. (3.35) for FLD.

For aluminium, storage modulus, $E_1 = 0.716 \times 10^{11}\ N/m^2$

We neglect its loss modulus; i.e., $\eta_1 = 0$.

Loss factor of the viscoelastic layer in longitudinal deformation, $\eta_2 = 0.5$

Therefore,

$$e_2 = \frac{E_2}{E_1} = \frac{2 \times 10^9}{0.716 \times 10^{11}} = 0.028$$

$$h_2 = \frac{H_2}{H_1} = \frac{2\ mm}{2\ mm} = 1$$

Applying Eq. (3.35) for FLD, the composite loss factor is given by

$$\eta \approx \frac{0.5 \times 0.028 \times 1 \left(3 + 6 \times 1 + 4 \times 1^2\right)}{1 + 0.028 \times 1 \left(3 + 6 \times 1 + 4 \times 1^2\right)} = \frac{0.5 \times 0.364}{1 + 0.364} = 0.133$$

### 3.5.2 Constrained layer damping (CLD) treatment

Typically, a CLD treatment shown in Fig. 3.12(b) is more effective than an FLD treatment shown in Fig. 3.12(a) in that the CLD treatment yields higher composite loss factor with lesser penalty in terms of additional weight. The shear deformation caused in a CLD beam is typically much more than the extensional deformation in an FLD beam. Therefore, the shear damping due to CLD treatment is much more than the extensional damping due to FLD treatment

A CLD treatment often consists of a thin stiff metal foil with a thin adhesive damping layer on one of its sides. This can easily be stuck or pasted on to the base plate to be damped. The overall or composite loss factor of the arrangement shown in Fig. 3.12 (b) is given by [1, 11]

$$\eta = \eta_{G2} Y g \Big/ \Big[ 1 + (2 + Y) g + (1 + Y)(1 + \eta_{G2}^2) g^2 \Big] \qquad (3.38)$$

where

$Y \approx 3.5 \ e_3 h_3$ is the stiffness parameter $\qquad (3.39)$

$\eta_{G2}$ = loss factor of the viscoelastic material (layer 2) in shear

$e_3 = E_3/E_1, \ h_3 = H_3/H_1$

g, the shear parameter, is given by [11]

$$g = \frac{G_2}{k_b^2 H_2} \left[ \frac{1}{E_1 H_1} + \frac{1}{E_3 H_3} \right] \qquad (3.40)$$

$G_2$ = storage shear modulus of the viscoelastic layer,

$k_b$ = the frequency-dependent flexural wave number of the composite structure, given by Eq. (2.53).

Equation (3.39) implies that the stiffer the constraining layer (metal foil), the higher the value of the stiffness parameter $Y$ and thence the composite loss factor $\eta$.

**Example 3.8** The 2-mm thick aluminium plate of Example 3.7 is now lined with a 1.0 mm thick aluminium foil with a thin (0.2 mm thick) adhesive damping layer on the inner side to provide constrained layer damping. If the shear modulus of the adhesive (damping) layer is $1.0 \times 10^7 (1 + j \ 1.4) \ N/m^2$, then evaluate the composite loss factor of the CLD plate at 25 Hz.

**Solution**

For aluminium, $E_1 = 0.7 \times 10^{11} \ N/m^2$ and $\rho_1 = 2700 \ kg/m^3$

Let us neglect its loss factor; i.e., $\eta_1 = 0$.

For the adhesive layer, $G_2 = 1.0 \times 10^7 \ N/m^2$ and $\eta_{G2} = 1.4$ (given)

Circular frequency, $\omega = 2\pi \times 25 = 157.1 \ rad/s$

$H_1 = 2$ *mm*, $H_3 = 1.0$ *m* and both layers are made of aluminium.

Therefore, $h_3 = H_3/H_1 = 1.0/2 = 0.5$, and $e_3 = E_3/E_1 = 1.0$.
As per Eq. (3.39), stiffness parameter, $Y \approx 3.5 \times 1.0 \times 0.5 = 1.75$

Thickness of the damping layer, $H_2 = 0.2$ *mm*.
As per Eq. (2.53), the bending wave number is given by

$$k_b = \left( \frac{\rho_1 A_1 \omega^2}{E_1 I_1} \right)^{1/4} = \left[ \frac{2700 \, b \times (2/1000)(157.1)^2}{0.7 \times 10^{11} \times b (2/1000)^3 / 12} \right]^{1/4}$$

$$= 7.3 \ m^{-1}$$

Substituting these values in Eq. (3.40) yields the following value for shear parameter:

$$g = \frac{1.0 \times 10^7}{(7.3)^2 (0.2/1000)} \left[ \frac{1}{0.7 \times 10^{11} \times (2/1000)} + \frac{1}{0.7 \times 10^{11} \times (1.0/1000)} \right]$$

$$= 20.1$$

Finally, making use of Eq. (3.38) we evaluate the composite loss factor:

$$\eta = \frac{1.4 \times 1.75 \times 20.1}{1 + (2 + 1.75) 20.1 + (1 + 1.75)(1 + 1.4^2)(20.1)^2}$$

Hence,
$$\eta = 0.0146$$

It may be noted that unlike the composite loss factor of FLD, which is independent of frequency, that of the CLD treatment is frequency dependent, even if properties $(G \ and \ \eta)$ of the damping layer are constant (independent of frequency). Thus, the effectiveness of the CLD treatment depends on the mode of vibration at the forcing frequency. The composite loss factor, $\eta$, will be maximum if the CLD is applied at and near the node where shear strain is maximum.

The quantitative results presented above are strictly for a simply supported beam. Qualitatively, however, they apply for designing CLD treatments for plates and sheet-metal components as well. In particular, near the clamped or fixed edges where the flexural strain is the maximum, CLD treatment will be most effective. Unlike FLD treatment, the CLD treatment need not be applied over the entire exposed surface. A modal analysis would help in selective or economic use of CLD treatment, particularly for narrow-band excitation.

In summary, the CLD treatment yields higher damping and imposes lesser penalty in terms of additional weight, as compared to the corresponding FLD treatment.

## 3.6   Active Vibration Control

An active vibration control (AVC) system reduces the vibration of a system by means of an actuator that applies variable control input. When mechanical power is supplied to an AVC system the actuator is said to be a fully active actuator. Another type of actuator is the so-called semi active actuator which is essentially a passive device which can store or dissipate energy. In this section, we restrict our discussion to the former; that is, a fully active actuator.

Active isolation actuators are generally of the following types:

(a) hydraulic actuators consisting of a reservoir, pump system and a fluidic circuit
(b) electromagnetic drives like voice coil actuators
(c) actuators incorporating active materials like piezoceramic, magnetostrictive and magneto-rheological materials.

Piezoceramic materials tend to exhibit relatively high force and low stroke. Therefore, they are often combined with a hydraulic or mechanical load-coupling mechanism to multiply motion at the expense of the applied force [12].

Active vibration control is effected by one or a combination of the following approaches:

(a) feedback control
(b) feedforward control

In general, active isolation of periodic or otherwise predictable excitations is achieved by means of the feedforward control approach. However, if the excitation is random or unpredictable, then the feedback control approach is preferred. The two types of controllers exhibit exactly the same performance for the case of sinusoidal disturbance [12].

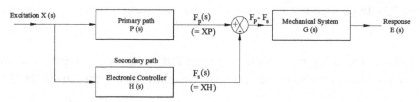

Fig. 3.13 Block diagram of a feedforward AVC system.

There is lot of literature on the feedback control of large flexible systems at relatively low frequencies. However, feedforward control is the preferred approach for active control of vibration and the associated sound radiation at audio frequencies and steady state excitation, which is of primary interest in this book. A block diagram of a feed forward control arrangement is shown in Fig. 3.13. Here, all the signals are represented by Laplace transform and the system dynamics are represented by transfer functions.

As is obvious from Fig. 3.13, the response E(s) is related to the excitation $X(s)$ as follows:

$$E(s) = X(s)\big[P(s) - H(s)\big]G(s) \tag{3.41}$$

Here, $F_p(s)$ and $F_s(s)$ are, respectively, the primary force and secondary force acting on the mechanical system.

Unlike feedback control systems, the feedforward control arrangements require a high degree of accuracy in magnitude and phase of the control system in order to obtain good cancellation. Therefore, most AVC systems make use of adaptive digital filters of the finite impulse response (FIR) type whose transfer function $H(s)$ is adapted via the error signal [13]. An example of such an AVC system is shown in Fig. 3.14 for control of flexural waves along a slender beam.

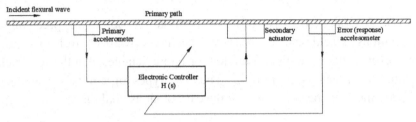

Fig. 3.14 Block diagram of an adaptive AVC system for control of flexural wave along a slender beam.

This type of active vibration control has been used, for example, to isolate vibrations from the gearbox traveling down the struts to the helicopter fuselage.

Active vibration control is particularly suited for engine mounts in an automobile. The disturbance vibrations that travel through the mounts to the chassis and thence to the passenger cabin of the vehicle are well correlated with the engine rotational speed. Therefore, the reference signal can be generated from the crankshaft pick-up. The error signal is obtained from accelerometer mounted on the chassis and/or a microphone located inside the passenger cabin. The controller may be located at a convenient location like the boot or trunk of the vehicle. The active engine mount enables control inputs to be applied to the vehicle in the load path of the standard rubber passive mounts by means of control actuators co-located and integrated with the passive mounts. However, there being several paths in a vehicle for vibration transmission, the AVC system requires multiple control actuators and error transducers for global reduction of vibration in the structure and noise in the cabin [15].

For the same reason, global control of sound from vibrating structures through on AVC approach would require multiple actuators, multiple microphones as error transducers, and a multiple-input multiple-output (MIMO) controller. Here too, we can use adaptive FIR filters and the least mean square (LMS) algorithm to adapt the FIR filters. A prior knowledge of dominant modes would help in that the controller could be designed to control the efficient radiating modes [13]. Thus, a finite-element modal analysis of the structure is often an essential component of an active structural acoustical control (ASAC) system.

An important application of AVC is in active vibration absorbers. Passive vibration absorbers suffer from the disadvantages of excessive stroke length and the added mass. Piezo-electric materials like ceramic (PZT) or a polymer (PVDF) act as a transformer between mechanical and electrical energy. Thus, the active vibration absorber replaces the flexure and mass with electronic analogs, and can achieve larger effective stroke length with less mass added to the system [14]. However, the active vibration absorbers (AVA) too suffer from certain disadvantages as follows.

(a) AVA needs separate power and electronics and hence is substantially costlier than its passive counterpart.
(b) AVA needs custom analog circuits or digital controllers.
(c) Unlike mechanical (passive) vibration absorbers, active vibration absorbers can lead to dynamical instability [14].

## References

1. Mallik, A. K., Principles of Vibration Control, Affiliated East-West Press, New Delhi, (1990).
2. Lindley, A. L. G. and Bishop, R. E. D., Some recent research on the balancing of large flexible rotors, Proceedings of the Institution of Mechanical Engineers, 177, pp. 811-26, (1963).
3. Munjal, M. L., A Rational Synthesis of Vibration Isolators, Journal of Sound and Vibration, 39(2), pp. 247-265, (1975).
4. Olson, M. F., Dynamical Analogies, Second Edition, Van Nostrand, Princeton, (1958).
5. Steffen, V. and Rade, D., Vibration Absorbers, pp. 9-26 in Braun, S.G. (Ed.) Encyclopedia of Vibration, Academic Press, San Diego, (2002).
6. Munjal, M. L., Acoustics of Ducts and Mufflers, John Wiley, New York, (1987).
7. Hayek, S. I., Vibration Transmission, pp. 1522-31 in Braun, S.G. (Ed.) Encyclopedia of Vibration, Academic Press, San Diego, (2002).
8. Cremer, L., Heckl, M. and Ungar, E. E., Structure-borne Sound, Springer-Verlag, New York, (1973).
9. Graff, K. F., Wave Motion in Elastic Solids, Ohio State University Press, Columbus, OH, (1975).

10. Ungar, E. E., Damping Materials, pp. 327-331 in Braun, S.G. (Ed.) Encyclopedia of Vibration, Academic Press, San Diego, (2002).
11. Ross, D., Ungar E. E. and Kerwin, E. M., (Jr.), Damping of plate flexural vibration by means of viscoelastic laminae, Section three, in J. E. Ruzicks (Ed.) Structural Damping, Pergamon Press, Oxford, (1960).
12. Griffin, S. and Sciulli, D., Active Isolation, pp. 46-48 in Braun, S.G. (Ed.) Encyclopedia of Vibration, Academic Press, San Diego, (2002).
13. Fuller, C. R., Feedforward control of Vibration, pp. 513-520 in Braun, J.G. (Ed.) Encyclopedia of Vibration, Academic Press, San Diego, (2002).
14. Agnes, G., Active absorbers, pp. 1-8 in Braun, S.G. (Ed.), Encyclopedia of Vibration, Academic Press, San Diego, (2002).

**Problems in Chapter 3**

**Problem 3.1** A diesel generator (DG) set running at 1500 RPM is to be isolated from the unbalanced forces by means of the anti-vibration mount (AVM) springs of Fig. 3.4 with stiffness of 912 N/mm (see Example 3.1). Masses of the engine, alternator and the base plate (inertia block) are 500 kg, 250 kg and 2,000 kg, respectively. How many of these AVM springs will be needed in order to limit transmissibility to 10%. Neglect the effect of damping.

[**Ans.: 6**]

**Problem 3.2** The unbalance in the rotor of a reciprocating compressor is 100 gm-cm. The mass of the compressor is 500 kg and it is operating at 1500 RPM, and is supported on rigid foundation through vibration isolators with static deflection of 10 mm. Calculate

(a) Amplitude of the unbalanced forces and the frequency of excitation
(b) Stiffness of the isolators and natural frequency of the system
(c) Transmissibility and amplitude of the force transmitted to the rigid foundation
(d) Transmissibility for a compliant foundation such that the static deflection is doubled

[**Ans.:** (*a*) **24.67 *N* at 25 *Hz*;** (*b*) **4.9×10⁵ *N/m*** and **4.98 *Hz*;**
(*c*) **0.041** and **1.02 *N*;** (*d*) **0.02** ]

**Problem 3.3** In order to arrest flanking transmission, an anechoic room, $5m \times 5m \times 5m$, is to be mounted on composite bonded rubber springs of the type shown in Fig. 3.3. All the six surfaces (floor, walls and ceiling) of the room are made of 200 mm thick RCC of density 2300 $kg / m^3$. Neglecting the weight of the acoustic wedges, evaluate the number of $0.5 \ m \times 0.5 \ m \times 0.05 \ m$ rubber pads in each of the 9 equally spaced springs so as to ensure natural frequency of 8 Hz. Assume elastic modulus of rubber to be $10^8 \ N / m^2$.

**[Ans.: 26]**

**Problem 3.4** A diesel generator (DG) set, weighing 1000 kg and running at 3000 RPM, is supported on rigid foundation through eight anti-vibration mounts of the type shown in Fig. 3.4, with $\alpha = 60^0$, $h = 5 \ cm$, $A = 50 \ cm^2$, damping ratio, $\zeta = 0.2$, $G = 3 \times 10^6 \ N/m^2$. Evaluate transmissibility of the system.

**[Ans.: 0.193]**

**Problem 3.5** A machine weighing 1000 kg is supported on a rigid foundation by means of vibration isolators of stiffness $10^6 \ N / m$. It is acted upon by unbalanced forces of $1000 \cos (150t) \ N$. Evaluate amplitudes of vibration displacement of the machine and the force transmitted to the rigid foundation as it is; i.e., without any DVA, and then design an undamped DVA (evaluate $k_a$ and $m_a$) in order to reduce vibration amplitude of the machine and the transmitted force to 25% of the original values for 2% variation in the forcing frequency.

**[Ans.:**
$X = 0.046 \ mm$, $F_t = 46.5N$, $m_a = 80 \ kg$ **and** $k_a = 1.8 \times 10^6 \ N/m$ **]**

**Problem 3.6** It is proposed to block transmission of longitudinal waves through a 3 *mm* diameter steel rod by means of a blocking mass $M$, as shown in Fig. 3.11. What would be the required mass for transmission loss of 10 dB at 100 Hz? What would be the *TL* due to this blocking mass for flexural waves?

**[Ans.: 2.71 kg; 3.0 dB]**

# Chapter 4

# Acoustics of Rooms, Partitions, Enclosures and Barriers

Automobiles, trains and aeroplanes operate in more or less open space. However, machines are often installed and operated in closed factories and sheds where reverberant or diffuse sound field coexists with the direct sound field, as shown in Fig. 4.1. The direct sound field is characterized by a spherical or hemi-spherical wave front and its sound pressure level or the intensity level is governed by the inverse square law, as discussed before in Chapter 1. The reverberant field, by contrast, is generated by multiple reflections from the room walls. In an ideal reverberant room (with zero or little absorption), the sound pressure field would be diffuse; a microphone moved around in such a room would show more or less the same SPL everywhere. In an anechoic room, on the other hand, there will be no reflection or reverberations at all. Such a room would support only the direct sound field, and therefore, an anechoic room is said to simulate free field like in open outdoors. All real rooms like living quarters, offices, factories and commercial establishments support direct field as well as diffuse field to varying degrees, depending upon the amount of acoustic absorption of the ceiling, walls, floor, windows and doors as well as the furniture, carpeting, curtains and human occupancy.

## 4.1 Sound Field in a Room

For a machine with sound power level at a particular frequency, $L_w(f)$, the sound pressure level $L_p(r, f)$ at a distance $r$ from the radiating surface or acoustical centre of the machine (see Fig. 4.1) is, in general, given by the following formula [1]:

$$L_p\left(r,f\right)=L_w\left(f\right)+10\ \log\left(\frac{Q(f)}{4\pi r^2}+\frac{4}{R(f)}\right) \tag{4.1}$$

Here, $Q(f)\big/4\pi r^2$ is the contribution from the direct field
$4/R(f)$ is the contribution from the diffuse field
$Q(f)$ is the directivity factor

$$Q(f)=Q_l\cdot Q_i(f) \tag{4.2}$$

where $Q_l$ is the locational directivity factor:

$$Q_l=2^{n_s} \tag{4.3}$$

$n_s$ is the number of surfaces touching at the machine or source.
Thus,

$Q_l=1$ for a source suspended midair away from all surfaces $\left(n_s=0\right)$;
 2 for a source located in the middle of the floor, away from all walls $\left(n_s=1\right)$;
 4 for a source located where the floor meets one of the walls $\left(n_s=2\right)$; and
 8 for a source located in a corner where the floor meets two walls $\left(n_s=3\right)$.

$Q_i(f)$ is the inherent directivity factor of the source as a function of frequency.

Frequency $f$ is often the centre frequency of an octave band or one-third octave band. $R(f)$ is the room constant [2]:

$$R(f)\ =\ \frac{S\bar{\alpha}(f)}{1-\bar{\alpha}(f)} \tag{4.4}$$

where $S$ is the total surface area of the room, including floor, ceiling, walls, furniture and human occupancy, and $\bar{\alpha}(f)$ is the overall (surface-averaged) acoustic power absorption coefficient

$$\bar{\alpha}(f)\ =\ \frac{\sum\limits_i S_i\alpha_i(f)}{S},\ \ S=\sum\limits_i S_i \tag{4.5}$$

where $S_i$ and $\alpha_i(f)$ are the area of the $i^{th}$ surface, and its absorption coefficient, respectively.

Incidentally, it may be noted from Eq. (4.1) above that in free field where there is no reflection, $\bar{\alpha}=1$, $R(f)$ tends to infinity, and then Eq. (4.1) yields the following direct field equation

$$L_p(r,f)_{direct\ field} = L_w(f) + 10\ \log\left(\frac{Q(f)}{4\pi r^2}\right) \tag{4.6}$$

where $Q(f)$ is given by Eq. (4.2) above.

In Eqs. (4.1) and (4.6) it is assumed that the microphone is in the far field. The criteria for the far field are complex [3]. For practical use, however, these may be simplified to

$$kr \geq 3, \quad r \geq 3l \tag{4.7 a-b}$$

where $k = \omega/c = 2\pi/\lambda$ is the wave number, and $l$ is the characteristic source dimension. For typical industrial sources of noise, the farfield criterion further simplifies to

$$r \geq 3l \tag{4.7b}$$

Equation (4.6) defines the inverse-square law for direct acoustic pressure field:

$$p_{rms}^2(r), \ I(r)\ \alpha\ \frac{1}{r^2} \tag{4.8}$$

provided $r$ is large enough to satisfy the farfield criteria (4.7). In terms of levels, the inverse square law becomes

$$L_p(r_1) - L_p(r_2) = 20\ \log(r_2/r_1), \ dB \tag{4.9}$$

As $20 \log 2 = 6$, Eq. (4.9) implies that SPL in the farfield decreases by 6 dB per doubling of the distance. Thus, the decrease is sharper nearer to the source but is comparatively milder as one moves farther away from the source [1-3].

In a room, industrial shed or workshop, as we move away from the source, the direct field term of Eq. (4.1) decreases progressively as per the inverse square law, so that near to the walls, it becomes negligible with respect to the reverberant or diffuse field term 4/R. Then Eq. (4.1) reduces to the diffuse field equation

$$L_p(f)_{diffuse\ field} = L_w(f) + 10\ \log\left(\frac{4}{R(f)}\right) \tag{4.10}$$

Making use of Eq. (4.4) for the room constant R, Eq. (4.10) becomes

$$L_p(f)_{diffuse\ field} = L_w(f) + 10 \log \left[ \frac{4(1 - \bar{\alpha}(f))}{S\bar{\alpha}(f)} \right] \tag{4.11}$$

Note that the diffuse field equation (4.11) is independent of the distance parameter $r$ while the direct field equation (4.6) is independent of the room surfaces.

It may be noted from Eq. (4.11) that as $\bar{\alpha}(f)$ tends to zero, the diffuse field SPL would predominate over the direct field practically everywhere. This is the basic principle of a reverberation room. All the six surfaces of a reverberation room are designed to be highly reflective, and all the three pairs of the opposite sides are constructed at an angle of 5 to 10 degrees so that standing wave pattern is eliminated. This is how a reverberation room is designed to ensure a diffuse pressure field.

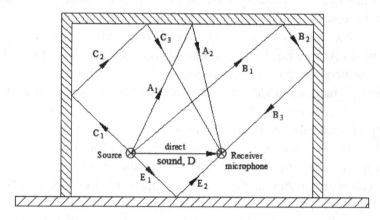

Fig. 4.1 Schematic of the direct sound field D superimposed on the reverberant sound field $A_1 - A_2$, $B_1 - B_2 - B_3$, $C_1 - C_2 - C_3$ and $E_1 - E_2$. The shortest distance between the source and the receiver microphone is denoted by r.

It may be noted from Eq. (4.1) that if $r$ tends to zero, the direct sound field would predominate over the diffuse field. This is why a whisper in the ear can be heard even in the noisiest of environments. Thus, in a reverberant room sound level meter would show the same SPL everywhere except very near the source. So, the microphone should not be located too near the sound source while making measurements in a reverberation room.

Effectiveness of a reverberation room to produce a diffuse field is measured in terms of Reverberation Time, $T_{60}$, which is defined as the time required for the average sound pressure level to decay by *60 dB* from its initial value (when the source is suddenly switched off). It is given by [2]

$$T_{60} = \frac{55.25V}{Sc\bar{\alpha}} \approx 0.161 \frac{V}{A} \tag{4.12}$$

where $V$ is the volume of the room, $S$ is the total area of all surfaces of the room, $c$ is sound speed, and A is the total absorption of the room in $m^2$:

$$A = S\bar{\alpha} = \sum_i S_i \alpha_i \tag{4.13}$$

In practice, however, a room will have many complex objects, apart from its bounding surfaces (floor, ceiling, walls, windows, doors and ventilators), and therefore, it is very difficult to evaluate the room absorption A from Eq. (4.13). Fortunately it is now very easy to measure the reverberation time of the room all over the frequency range of interest. In fact, the random-incidence absorption coefficient for acoustic materials is now measured in a reverberation room as a standard practice [4, 5]. The procedure in brief is as follows.

A blanket or mat of the test material is placed in the centre of the floor in a reverberation room and the reverberation time is measured with and without the material. Then, repeated application of Eqs. (4.12) and (4.13) yields [4, 6]

$$A \equiv S \, \alpha \approx 0.161 \, V \left[ \frac{1}{T_{60}} - \frac{1}{T_{60}'} \right] \tag{4.14}$$

where $T_{60}$ and $T_{60}'$ are values of the reverberation time measured with and without the test material blanket or object, respectively, in the frequency band of interest. Thus absorption coefficient $\alpha$ of an acoustic material or absorption A of a test object can readily be evaluated experimentally.

Absorption coefficient of different types of walls, ceilings, furnishings, linings and panels at centre frequencies of different octave bands, measured in specially designed reverberation rooms, are listed in handbooks —— see, for example, references [1, 6, and 7]. Some of them are reproduced here from Ref. [6] in Table 4.1.

As the pressure field in a reverberation room is more or less uniform throughout the room, Eq. (4.11) presents a convenient method of evaluating the total sound power level of a machine. Microphone at the end of the arm of a rotating boom is used to measure the space-average SPL in the frequency band of interest. Absorption A of the reverberation room is evaluated from the measured reverberation time and used in Eq. (4.11). This is in fact a standard technique for measurement of sound power level of small portable machines for purposes of labeling as well as environmental impact assessment [4, 8].

**Example 4.1** The floor, walls and ceiling of a 15 m×10 m×5 m (*high*) room are made of varnished wood joists, bricks, and 13 mm suspended mineral tiles, respectively. Calculate the average absorption coefficient and reverberation time of the room in the 500 Hz frequency band when the floor is:

(a) unoccupied and unfurnished
(b) unoccupied but furnished with well-upholstered seats
(c) 100% occupied

**Solution**

Volume of the room = $15 \times 10 \times 5 = 750$ m$^3$
Area of the floor     = $15 \times 10 = 150$ m$^2$
Area of the ceiling  = $15 \times 10 = 150$ m$^2$
Area of the walls    = $2(15 + 10) \times 5 = 250$ m$^2$

Total surface area of the room, $S = \Sigma S_i = 150 + 150 + 250 = 550$ m$^2$

Referring to Table 4.1, values of the sound absorption coefficient at 500 Hz for different surfaces are as follows:

Bare varnished wood joists floor: 0.1
Floor unoccupied but furnished with well upholstered seats: 0.77
100% occupied floor: 0.85
Brick walls: 0.03
13 mm suspended mineral tile ceiling: 0.65

Substituting these data on areas $S_i$ and sound absorption coefficient $\alpha_i$ in Eq. (4.5) yields the following values of the surface-averaged $\bar{\alpha}$ at 500 Hz:

$$\bar{\alpha}(bare\ room) = \frac{150 \times 0.1 + 250 \times 0.03 + 150 \times 0.65}{550} = 0.218$$

$$\bar{\alpha}(upholstered\ seats) = \frac{150 \times 0.77 + 250 \times 0.03 + 150 \times 0.65}{550} = 0.4$$

$$\bar{\alpha}(occupied\ room) = \frac{150 \times 0.85 + 250 \times 0.03 + 150 \times 0.65}{550} = 0.423$$

Substituting each of these three values of $\bar{\alpha}$ in Eq. (4.12) yields the corresponding values of the Reverberation Time, $T_{60}$. Thus,

$$T_{60}(bare\ room) = \frac{0.161 \times 750}{550 \times 0.218} = 1.0\ s$$

$$T_{60}(upholstered\ seats) = \frac{0.161 \times 750}{550 \times 0.4} = 0.55\ s$$

$$T_{60}(occupied\ room) = \frac{0.161 \times 750}{550 \times 0.423} = 0.52\ s$$

**Example 4.2** A noisy machine is located in the middle of the floor of the bare room of Example 4.1 ($S = 550\ m^2$, $\bar{\alpha} = 0.218$). Evaluate the distance (reckoned from the centre of the machine) at which the direct field would be as strong as the diffuse field.

**Solution**

$$L_p(r) = L_w + 10\ \log\left(\frac{Q}{4\pi r^2} + \frac{4}{R}\right)$$

As the machine is touching one surface (the floor), $Q = 2^1 = 2$

Room constant, $R = \dfrac{S\bar{\alpha}}{1-\bar{\alpha}} = \dfrac{550 \times 0.218}{1 - 0.218} = 153.3$ m$^2$

Contribution of the direct sound would be equal to the reverberation sound (diffuse field) if

$$\frac{Q}{4\pi r^2} = \frac{4}{R}$$

This would happen at

$$r = \frac{1}{4}\left(\frac{QR}{\pi}\right)^{1/2} = \frac{1}{4}\left(\frac{2 \times 153.3}{\pi}\right)^{1/2} = 2.47 \text{ m}$$

Within a radius of 2.47 m, the direct sound would dominate, and outside this radius, the diffuse field would dominate.

Note that this neutral distance is a function of $\bar{\alpha}$, which in turn is a function of frequency (500 Hz in Example. 4.1)

## 4.2 Acoustics of a Partition Wall

Often a noisy source in a room is separated from the receiver by means of a partition wall. When this wall does not extend upto the ceiling, it is called a barrier. Acoustics of a barrier is discussed later in this chapter. A partition wall reflects a part of the incident acoustic power or energy back to the source side of the room, absorbs a little of it and lets the rest pass through to the other side, as shown in Fig. 4.2. Then,

$$W_i = W_r + W_a + W_t \qquad (4.15)$$

Transmission coefficient, $\tau = \dfrac{W_t}{W_i}$ \qquad (4.16)

Reflection coefficient, $R = \dfrac{W_r}{W_i}$ \qquad (4.17)

Table 4.1 Sound absorption coefficient for common materials, objects and surfaces [6].

| Material/Surface/Object | Octave band centre frequency (Hz) | | | | | |
|---|---|---|---|---|---|---|
| | 125 | 250 | 500 | 1000 | 2000 | 4000 |
| Unoccupied average well-upholstered seating areas | 0.44 | 0.60 | 0.77 | 0.84 | 0.82 | 0.70 |
| Unoccupied metal or wood seats | 0.15 | 0.19 | 0.22 | 0.39 | 0.38 | 0.30 |
| 100% occupied audience (orchestra and chorus areas) – upholstered seats | 0.52 | 0.68 | 0.85 | 0.97 | 0.93 | 0.85 |
| Audience, per person seated $S\bar{\alpha}(m^2)$ | 0.23 | 0.37 | 0.44 | 0.45 | 0.45 | 0.45 |
| Wooden chairs – 100% occupied | 0.60 | 0.74 | 0.88 | 0.96 | 0.93 | 0.85 |
| Fibre-glass or rockwool blanket 24 kg/m$^3$, 50 mm thick | 0.27 | 0.54 | 0.94 | 1.0 | 1.0 | 1.0 |
| Fibre-glass or rockwool blanket 24 kg/m$^3$, 100 mm thick | 0.46 | 1.0 | 1.0 | 1.0 | 1.0 | 1.0 |
| Fibre-glass or rockwool blanket 48 kg/m$^3$, 100 mm thick | 0.65 | 1.0 | 1.0 | 1.0 | 1.0 | 1.0 |
| Polyurethane foam, 27 kg/m$^3$, 15 mm thick | 0.08 | 0.22 | 0.55 | 0.70 | 0.85 | 0.75 |
| Concrete or terrazzo floor | 0.01 | 0.01 | 0.01 | 0.02 | 0.02 | 0.02 |
| Varnished wood joist floor | 0.15 | 0.12 | 0.10 | 0.07 | 0.06 | 0.07 |
| Carpet, 5 mm thick, on hard floor | 0.02 | 0.03 | 0.05 | 0.10 | 0.30 | 0.50 |
| Glazed tile/marble | 0.01 | 0.01 | 0.01 | 0.01 | 0.02 | 0.02 |
| Hard surfaces (brick walls, plaster, hard floors, etc.) | 0.02 | 0.02 | 0.03 | 0.03 | 0.04 | 0.05 |
| Gypsum board on 50 x 100 mm studs. | 0.29 | 0.10 | 0.05 | 0.04 | 0.07 | 0.09 |
| Solid timber door | 0.14 | 0.10 | 0.06 | 0.08 | 0.10 | 0.10 |
| 13mm mineral tile direct fixed to ceiling slab | 0.10 | 0.25 | 0.70 | 0.85 | 0.70 | 0.60 |
| 13mm mineral tile suspended 500 mm below ceiling | 0.75 | 0.70 | 0.65 | 0.85 | 0.85 | 0.90 |
| Light velour, 338 g/m$^2$ curtain hung on the wall | 0.03 | 0.04 | 0.11 | 0.17 | 0.24 | 0.35 |
| hung in folds on wall | 0.05 | 0.15 | 0.35 | 0.40 | 0.50 | 0.50 |
| Heavy velour, 610 g/m$^2$ curtain draped to half area | 0.14 | 0.35 | 0.55 | 0.72 | 0.70 | 0.65 |
| Ordinary Window Glass | 0.35 | 0.25 | 0.18 | 0.12 | 0.07 | 0.04 |
| Water (surface of pool) | 0.01 | 0.01 | 0.01 | 0.015 | 0.02 | 0.03 |

$$\text{Absorption coefficient, } \alpha = \frac{W_a}{W_i} \qquad (4.18)$$

$$\text{Transmission loss, } TL = 10 \log\left(\frac{W_i}{W_t}\right) = -10 \log(\tau) \qquad (4.19)$$

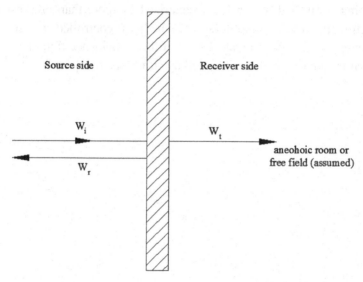

Fig. 4.2 Schematic of a partition wall for normal incidence transmission loss: $TL = 10 \log(W_i/W_t)$.

For most partition walls, absorption is negligible; i.e., $\alpha \ll 1$, unless the wall is lined with an acoustically absorptive layer. A major portion of the incident power is reflected back to the source side owing to the impedance mismatch between the air medium (on either side of the wall) and the material of the wall. For typical partition walls, lumped impedance of the wall, $Z$, normalized with respect to the characteristic impedance of the surrounding medium $(\rho_0 c_0)$ (mostly, air) plays a primary role. It can be shown that transmission loss of a partition wall is given by

$$TL = 20 \log\left|1 + \frac{Z}{2\rho_0 c_0}\right| \qquad (4.20)$$

Typically, impedance consists of a stiffness component, mass component and structural damping, and therefore a typical TL curve when plotted against frequency (on a log scale) has a stiffness controlled region (O-A) and mass controlled region (B-C) separated by a damping controlled region (A-B), as shown in Fig. 4.3. D denotes a Coincidence dip and the region beyond C is called the Coincidence region. The damping controlled region is characterized by more than one resonance troughs, the lowest amplitude of which is controlled by structural damping of the partition wall. Depth of the coincidence dip, or TL at the critical frequency, is also controlled by structured damping.

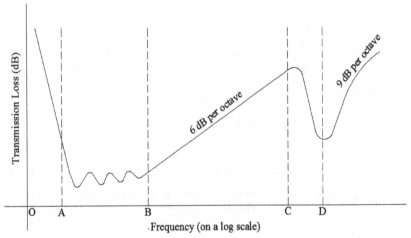

Fig. 4.3 Typical TL curve for a partition wall, showing the stiffness-controlled region, O – A, the damping-controlled region, A – B, the mass-controlled region, B – C, the coincidence (or, critical frequency) dip, D, and the coincidence-controlled region, beyond C.

For practical purposes, the mass-controlled region is most important, where

$$Z \simeq j\omega m, \quad m = \rho h, \quad \omega = 2\pi f \qquad (4.21)$$

Here, $m$ is surface density, defined as mass per unit area of the wall,
$\rho$ is density of the partition wall material, and
$h$ is thickness of the wall.

Substituting Eq. (4.21) in Eq. (4.20) yields the following formula for normal incidence TL:

$$TL = 10 \, \log\left\{1+\left(\frac{\omega \rho h}{2\rho_0 c_0}\right)^2\right\} \simeq 20 \, \log\left\{\frac{\pi f \rho h}{\rho_0 c_0}\right\} \quad \text{dB} \qquad (4.22)$$

Equation (4.22) applies for normal incidence. For random or field incidence we need to subtract 5.5 dB for one-third octave band measurements and 4 dB for octave band measurements.

It may be noted from Eq. (4.22) that in the mass-controlled region, TL increases by 6 dB per octave. Also, TL would increase by 6 dB if $\rho$, $h$ or mass density $m$ were doubled. In other words, thicker or denser partition walls would provide better acoustic insulation, and a given wall would provide more TL at higher frequencies; in fact, 6 dB additional TL per octave.

The phenomenon of coincidence occurs when the following relationship is satisfied for an incident wave striking an infinite wall or panel (see Fig. 4.4):

$$\lambda_b = \frac{\lambda_0}{\sin\theta}, \quad \lambda_0 = \frac{2\pi}{k_0} = \frac{c_0}{f} \qquad (4.23)$$

where $\lambda_b$ is the bending or flexural wave length,

$$\lambda_b = \frac{2\pi}{k_b}, \quad k_b = \left(\frac{m\omega^2}{EI}\right)^{1/4} \qquad (4.24)$$

$m$ is the surface density of the wall;

$EI$ is the flexural rigidity of the wall;

$\theta$ is the angle of incidence; and

$k_b$ is the flexural wave number.

It is obvious from the coincidence relation (4.23) that coincidence would occur at only those frequencies that satisfy the inequality

$$\lambda_b \geq \lambda_0 \qquad (4.25)$$

This inequality would be satisfied at all frequencies exceeding the coincidence frequency given by:

$$f_c = \frac{c_0^2}{2\pi}\left(\frac{m}{EI}\right)^{1/2} \tag{4.26}$$

Above this critical frequency, wavelength of the flexural or bending wave in the wall will become equal to the projection of the wavelength of the incident wave upon the wall at an appropriate angle $\theta$, as shown in Fig. 4.4.

Incidentally, Eq. (4.26) is particularly significant inasmuch as it represents the cut-off frequency below which a vibrating wall (or plate) would not radiate any sound, and conversely, an incident wave would not be able to set the wall (or plate) into vibration. This important physical principle is of great use in control of noise from vibrating bodies, as will become clear later in Chapter 6.

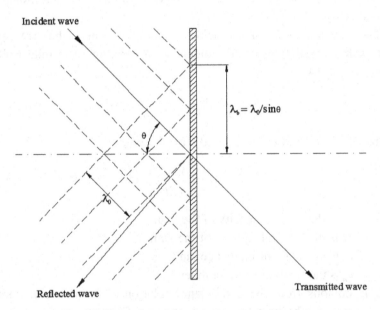

Fig. 4.4 Illustration of the phenomenon of coincidence.

In practice, TL of partition walls, including panels, masonry walls, stud partitions, glazed windows, doors and floors, is measured in a reverberation suite for random incidence in different octave band centre frequencies. For some common partition walls, the measured values of the random-incidence transmission loss are listed in Table 4.2. These are extracted from Ref. [6].

It may be noted from Table 4.2 that the random-incidence TL of common structures and materials does not necessarily follow the shape of Fig. 4.3, not even the well-defined mass law. Nevertheless, Fig. 4.3 does serve as a useful guideline.

In particular, it is obvious that a double-glaze window with substantial airgap $(\geq 50 \text{ mm})$ inbetween  yields substantially more TL than a single-glaze window of double the thickness, because of two pairs of impedance mismatch in the case of a double-glaze window, or for that matter, double-skin door, double masonry walls, stud partitions, etc.

It may also be noted that it is advisable to fill up the whole or a part of the intervening airgap with an absorptive material.

**Example 4.3**  Making use of the mass law, calculate the normal incidence *TL* of a 16 g galvanized steel sheet at mid-frequencies of the octave bands of 63 Hz to 8000 Hz, and compare the same with the corresponding random-incidence values from Table 4.2.

**Solution**

Thickness of 16 g sheet, $h = 1.6$ mm
Mass density of steel, $\rho = 7800$ kg / m$^3$
Therefore, surface density of 16 g steel sheet,
$$M = \rho h = (1.6/1000) \times 7800 \ = 12.48 \ \text{kg}/\text{m}^2$$

Characteristic impedance of air at $25^o C$, $\rho_0 c_0 = 410$ kg / (m$^2$s)

As per the mass law (Eq. 4.22), the normal-incidence TL of a sheet is given by

$$TL = 20 \ \log \left( \frac{\pi f \rho h}{\rho_0 c_0} \right) = 20 \ \log \left( \frac{\pi f \times 7800 \times 1.6}{410 \times 1000} \right)$$

or

$$TL = -20.4 + 20 \ \log (f) \quad \text{dB}$$

Values calculated from this relation are listed below in the second row.

| Frequency (Hz) | 63 | 125 | 250 | 500 | 1k | 2k | 4k | 8K |
|---|---|---|---|---|---|---|---|---|
| Normal Incidence, TL (as per mass law) | 15.5 | 21.5 | 27.5 | 33.5 | 39.5 | 45.5 | 51.5 | 57.5 |
| Random-Incidence, TL (measured; Table 4.2) | 9 | 14 | 21 | 27 | 32 | 37 | 43 | 42 |

It may be noted from the table that the normal-incidence mass law considerably overpredicts TL as compared to the measured values of *TL* for random incidence. This observation is very significant for designers of acoustic walls and enclosures. They should make use of the measured values of the random-incidence *TL* rather than the normal-incidence mass law.

## 4.3   Design of Acoustic Enclosures

Performance of an acoustic enclosure is measured in terms of its insertion loss. It is defined as reduction of SPL at the receiver due to location of the source (machine) in an acoustic enclosure. It is given by the following, rather approximate, relationship:

$$IL = 10 \log\left(\bar{\alpha}/\bar{\tau}\right) = TL + 10 \log \bar{\alpha} \qquad (4.27)$$

Here $\bar{\tau}$ and TL are the transmission coefficient and transmission loss of the impervious (often metallic) layer of the enclosure walls, and $\bar{\alpha}$ is the random-incidence absorption coefficient of the absorptive lining on the inner or source side of the enclosure walls (see Eq. 4.5).

It may be noted from Eq. (4.27) that IL of the enclosure is always less than TL of the walls because $\alpha$ is always less than unity. It also indicates importance of the absorptive lining of all the inner surfaces of the acoustic enclosure. In the absence of the lining, the diffuse pressure field inside would be very high and therefore IL would be very small despite metallic walls of high surface density.

The surface-averaged transmission coefficient $\bar{\tau}$ is given by an equation similar to Eq. (4.5) for $\bar{\alpha}$. Thus,

Table 4.2 Measured values of the random-incidence TL of typical partition walls [6].

| Panel Construction | Thickness (mm) | Surface density (kg/m²) | Octave band centre frequency (Hz) | | | | | | | |
|---|---|---|---|---|---|---|---|---|---|---|
| | | | 63 | 125 | 250 | 500 | 1k | 2k | 4k | 8k |
| **Panel of sheet materials** | | | | | | | | | | |
| 1.5 mm lead sheet | 1.5 | 17 | 22 | 28 | 32 | 33 | 32 | 32 | 33 | 36 |
| 6 mm steel plate | 6 | 50 | – | 27 | 35 | 41 | 39 | 39 | 46 | – |
| 22 g galvanized steel sheet | 0.75 | 6 | 3 | 8 | 14 | 20 | 23 | 26 | 27 | 35 |
| 16 g galvanized steel sheet | 1.6 | 13 | 9 | 14 | 21 | 27 | 32 | 37 | 43 | 42 |
| Fibre board on wood frame work | 12 | 4 | 10 | 12 | 16 | 20 | 24 | 30 | 31 | 36 |
| Plastic board sheets on wood frame work | 9 | 7 | 9 | 15 | 20 | 24 | 29 | 32 | 35 | 38 |
| Woodwork slabs, unplastered | 25 | 19 | 0 | 0 | 2 | 6 | 6 | 8 | 8 | 10 |
| Woodwork slabs, plastered (12 mm on each face) | 50 | 75 | 18 | 23 | 27 | 30 | 32 | 36 | 39 | 43 |
| Plywood | 6 | 3.5 | – | 17 | 15 | 20 | 24 | 28 | 27 | – |
| Lead vinyl curtains | 2 | 4.9 | – | 15 | 19 | 21 | 28 | 33 | 37 | – |
| **Single masonry walls** | | | | | | | | | | |
| Single leaf brick, plastered on both sides | 125 | 240 | 30 | 36 | 37 | 40 | 46 | 54 | 57 | 59 |
| Single leaf brick, plastered on both sides | 255 | 480 | 34 | 41 | 45 | 48 | 56 | 65 | 69 | 72 |
| Single leaf brick, plastered on both sides | 360 | 720 | 36 | 44 | 43 | 49 | 57 | 66 | 70 | 72 |
| Hollow cinder concrete blocks, painted (cement base paint) | 100 | 75 | 22 | 30 | 34 | 40 | 50 | 50 | 52 | 53 |
| Hollow cinder concrete blocks, unpainted | 100 | 75 | 22 | 27 | 32 | 32 | 40 | 41 | 45 | 48 |
| Aerated concrete blocks | 100 | 50 | – | 34 | 35 | 30 | 37 | 45 | 50 | – |
| Aerated concrete blocks | 150 | 75 | – | 31 | 35 | 37 | 44 | 50 | 55 | – |

| Panel Construction | Thickness (mm) | Surface density (kg/m²) | Octave band centre frequency (Hz) | | | | | | | |
|---|---|---|---|---|---|---|---|---|---|---|
| | | | 63 | 125 | 250 | 500 | 1k | 2k | 4k | 8k |
| **Stud Partitions** | | | | | | | | | | |
| 50 mm x 100 mm studs, 12 mm board both sides | 125 | 19 | 12 | 16 | 22 | 28 | 38 | 50 | 52 | 55 |
| 50 mm x 100 mm studs, 9 mm plasterboard and 12 mm plaster coat both sides | 142 | 60 | 20 | 25 | 28 | 34 | 47 | 39 | 50 | 56 |
| Gypsum wall with steel studs and 16 mm-thick panels each side | | | | | | | | | | |
| Empty cavity, 45 mm wide | 75 | 26 | – | 20 | 28 | 36 | 41 | 40 | 47 | – |
| Cavity, 45 mm wide, filled with fiberglass | 75 | 30 | – | 27 | 39 | 46 | 43 | 47 | 52 | – |
| Gypsum wall, 16 mm leaves, 200 mm cavity with no sound absorbing material and no studs | 240 | 23 | – | 33 | 39 | 50 | 64 | 51 | 59 | – |
| As above with 88 mm sound absorbing material | 240 | 26 | – | 42 | 56 | 68 | 74 | 70 | 73 | – |
| **Single glazed windows** | | | | | | | | | | |
| Single glass in heavy frame | 6 | 15 | 17 | 11 | 24 | 28 | 32 | 27 | 35 | 39 |
| Single glass in heavy frame | 8 | 20 | 18 | 18 | 25 | 31 | 32 | 28 | 36 | 39 |
| Laminated glass | 13 | 32 | – | 23 | 31 | 38 | 40 | 47 | 52 | 57 |

| Panel Construction | Thickness (mm) | Surface density (kg/m²) | Octave band centre frequency (Hz) | | | | | | | |
|---|---|---|---|---|---|---|---|---|---|---|
| | | | 63 | 125 | 250 | 500 | 1k | 2k | 4k | 8k |
| **Doubled glazed window** | | | | | | | | | | |
| 2.44 mm panes, 7 mm cavity | 12 | 15 | 15 | 22 | 16 | 20 | 29 | 31 | 27 | 30 |
| 9 mm panes in separate frames, 50 mm cavity | 62 | 34 | 18 | 25 | 29 | 34 | 41 | 45 | 53 | 50 |
| 6 mm glass panes in separate frames, 100 mm cavity | 112 | 34 | 20 | 28 | 32 | 38 | 45 | 45 | 53 | 50 |
| 6 mm and 8 mm glass, 100 mm cavity | 115 | 40 | – | 35 | 47 | 53 | 55 | 50 | 55 | – |
| **Doors** | | | | | | | | | | |
| Flush panel, hollow core, normal cracks as usually hung | 43 | 9 | 1 | 12 | 13 | 14 | 16 | 18 | 24 | 26 |
| Solid hardwood, normal cracks as usually hung | 43 | 28 | 13 | 17 | 21 | 26 | 29 | 31 | 34 | 32 |
| Typical proprietary "acoustic" door, double heavy sheet steel skin, absorbent in air space, and seals in heavy steel frame | 100 | – | 37 | 36 | 39 | 44 | 49 | 54 | 57 | 60 |
| Hardwood door | 54 | 20 | – | 20 | 25 | 22 | 27 | 31 | 35 | – |

$$\bar{\tau}(f) = \frac{\sum S_i \tau_i (f)}{S}, \quad S = \sum_i S_i \tag{4.28}$$

As $\tau$ for openings or leaks is nearly unity, leaks could seriously compromise $\bar{\tau}$ and hence TL. For example, access openings or leaks of 1% would limit TL to 20 dB irrespective of the surface density of the walls of the enclosure. Similarly, access openings of 10% would limit TL to 10 dB. This indicates how ineffective could partial enclosures be, except at high frequencies where one could effectively make use of directivity of the high-frequency sounds. Partial enclosures would yield low IL values for low-frequency sounds, which tend to propagate spherically, easily skirting around the openings.

For practical purposes, Eq. (4.27) may be rewritten as

$$IL = TL - C, \quad C \equiv -10 \log \bar{\alpha} \tag{4.29}$$

The experimentally determined values of C as functions of frequency for different types of enclosures are listed in Table 4.3 which is reproduced from Bies and Hansen [6].

Table 4.3 Values of constant C (dB) to account for enclosure internal acoustic conditions.

| Enclosure internal acoustic conditions* | Octave band centre frequency (Hz) | | | | | | | |
|---|---|---|---|---|---|---|---|---|
| | 63 | 125 | 250 | 500 | 1000 | 2000 | 4000 | 8000 |
| Live | 18 | 16 | 15 | 14 | 12 | 12 | 12 | 12 |
| Fairly live | 13 | 12 | 11 | 12 | 12 | 12 | 12 | 12 |
| Average | 13 | 11 | 9 | 7 | 5 | 4 | 3 | 3 |
| Dead | 11 | 9 | 7 | 6 | 5 | 4 | 3 | 3 |

* Use the following criteria to determine the appropriate acoustic conditions inside the enclosure:

Live        : All enclosure surfaces and machine surfaces hard and rigid

Fairly live : All surfaces generally hard but some panel construction (sheet metal or wood)

Average    : Enclosure internal surfaces covered with sound-absorptive material, and machine Surfaces hard and rigid

Dead       : As for "Average", but machine surface mainly of panels

In most application, the enclosure's internal surfaces are covered with a sound-absorptive material, whereas the machine surfaces are hard and rigid. Therefore, the 'average' row (the third row) of Table 4.3 would apply for constant C. Lining the machine surfaces with acoustic panels (see the 'dead' or the fourth row) gives only a marginal improvement in C and thence IL, and is therefore not cost effective.

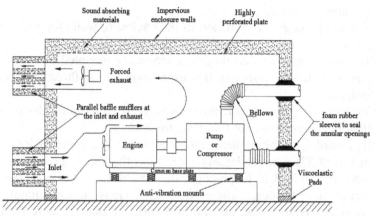

Fig. 4.5 Schematic of an acoustical enclosure used for noise control of an engine-driver pump or compressor [9].

Figure 4.5 shows schematic of an acoustic enclosure [9]. The following features of the acoustic enclosure may be noted:

1. All the available surfaces of walls and ceiling are lined with absorptive materials.
2. Parallel baffle mufflers are provided with sufficient openings for inlet and exhaust of cooling air.
3. The intake air is properly guided onto the hot surfaces of the machine. When ventilation for heat removal is required but the heat load is not large, then natural ventilation with silenced air inlets low down close to the floor and silenced outlets at a greater height well above the floor, will be adequate.
4. Generally, forced ventilation is needed, and indeed provided, by means of booster fans at the inlet and/or exhaust points.
5. If forced ventilation is needed to avoid excessive heat buildup in the enclosure, then the approximate amount of airflow needed can be determined by $\rho C_p V = H / \Delta T$, where V is the volume flow rate of the cooling air required to limit the steady state temperature increase inside the enclosure to $\Delta T^\circ C$, $\rho$ and $C_p$ are respectively

the mass density and the specific heat at constant pressure of the cooling air, and H is the heat input to the enclosure in watts.

6. Flexible connectors are used to isolate the enclosure walls from the machine vibrations.
7. Sealed openings are provided in the enclosure walls for any service pipes.
8. The noisy machine and its driver motor are mounted on a common platform, which rests on the foundation block through vibration isolators.
9. Viscoelastic pads support the enclosure all around in order to arrest flanking transmission of the structure-borne sound.
10. It is worth mentioning here that for a practical acoustic enclosure, *TL* in Eq. (4.27) is considerably more than the values given for a flat plate in Table 4.2. In practice, the enclosure is made of box-like rectangular frames, where the impervious layer (plate) gets considerably stiffened, and *TL* of a stiffened plate is more than that of a flat plate (of the same surface density) by several decibels. Some of this advantage may, however, be lost in acoustic leaks at the inter-frame joints if sufficient care is not exercised in assembling of the enclosure walls.

Figure 4.6 shows different types of penetrations (for cooling) lined with absorptive materials [9]. Theory and design of parallel baffle mufflers as well as lined labyrinth ducts will be discussed in the next chapter.

### 4.4  Noise Reduction of a Partition Wall and Enclosure

Often, one needs to relate SPL outside $\left(L_{p2}\right)$ with the SPL inside $\left(L_{p1}\right)$ an existing enclosure or room with access openings and ventilation louvers. Difference of the two SPLs is called Noise Reduction, NR. Thus,

$$NR = L_{p1} - L_{p2} \qquad (4.30)$$

Prediction of NR is particularly needed in an EIA (Environmental Impact Assessment) exercise. Towards this end, some useful relationships are reproduced below.

With reference to Fig. 4.7, NR is given by [1]

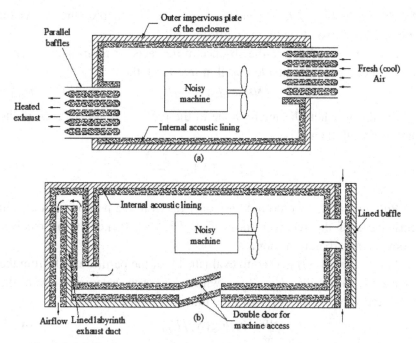

Fig. 4.6 Schematic plans of two silenced ventilation opening arrangements
(a) parallel baffles and (b) lined baffles with double-door access [9].

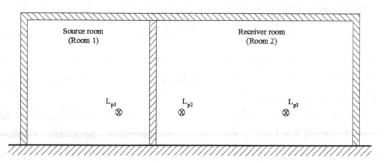

Fig. 4.7 Schematic of partition wall for illustration of Noise Reduction (NR).

$$NR = L_{p1} - L_{p2} = TL - 10 \, \log\left( \frac{1}{4} + \frac{S_w}{R_2} \right) \, \text{dB} \qquad (4.31)$$

where $TL$ and $S_w$ are transmission loss and area of the partition wall, respectively, and $R_2$ is the room constant of the receiver room (Room 2).

As shown in Fig. 4.7, $L_{p1}$ and $L_{p2}$, are the sound pressure levels on either side of, and close to, the partition wall.

It may be noted that in the absence of the receiver room; that is, when $L_{p2}$ is measured just outside a wall in free field, then

$$NR = TL + 6 \quad dB \tag{4.32}$$

Similarly, when the microphone in the receiver room is far from the partition wall, then [1],

$$L_{p1} - L_{p3} = TL - 10 \ \log\left(\frac{S_w}{R_2}\right) \tag{4.33}$$

Equation (4.33) would hold not only far from the partition but also in most parts of the receiver room if $S_w/R_2 \gg 1/4$; i.e., if the receiver room is a reverberation room.

In fact, Eq. (4.33) is used to evaluate TL of the partition wall from the measured values of $L_{p1} - L_{p3}$, with $R_2$ being calculated from the measured value of the reverberation time of Room 2. Thus,

$$TL = NR + 10 \ \log\left(S_w/R_2\right) \tag{4.34}$$

Making use of energy balance it can be shown that $L_{pi}$ and $L_{po}$, the sound pressure level inside and immediately outside the enclosure, respectively, are related as follows:

$$NR = L_{pi} - L_{po} = TL - C = IL \tag{4.35}$$

where constant $C$ is given in Table 4.3 and TL is the transmission loss of the enclosure walls. Thus, noise reduction of an acoustic enclosure is equal to its insertion loss.

To predict the average SPL on a hypothetical parallelepiped surface, say at 1 m $(d = 1 \text{ m})$ around a rectangular acoustic enclosure $(l \times b \times h)$ with known insertion loss IL in a room with room constant R in the frequency band of interest, we make use of the following relationship:

$$L_p = L_w - IL + 10 \ \log\left\{\frac{1}{S_m} + \frac{4}{R}\right\} \tag{4.36}$$

where, as shown in Fig. 4.8, area of the measurement surface, $S_m$, is given by

$$S_m = 2(l+2d)(h+d) + 2(b+2d)(h+d) + (l+2d)(b+2d) \qquad (4.37)$$

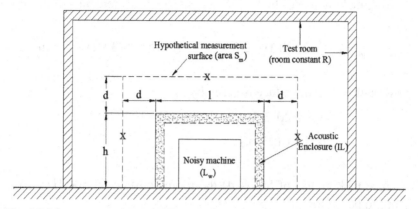

Fig. 4.8 Measurement of average SPL on a hypothetical parallelepiped surface inside a large room.

**Example 4.4** A diesel generator (DG) set with sound power level of 120 dB in the 500-Hz octave band is located in the middle of the floor of a 20 m×15 m×6 m industrial shed with average absorption coefficient of 0.1 in the same frequency band. The DG set is provided with a 4 m×3 m×3 m acoustic enclosure made of a 1.6 mm thick GI plate lined on the inside with 100 mm thick layer of mineral wool. Evaluate the average SPL at 1 m around the enclosure surface.

**Solution**

We make use of Eq. (4.36) along with Eq. (4.37) and Fig. 4.8 for the measurement surface area $S_m$, Eq. (4.29) along with Tables 4.2 and 4.3 for the enclosure *IL*, and Eq. (4.4) along with Table 4.1 for the room constant *R*. $L_w = 120 \, dB$ (given).

Referring to Table 4.2,
    *TL* of a 1.6 mm thick GI plate (at 500 Hz) = 27 dB
Referring to Table 4.3,
    for average acoustic lining, C (at 500 Hz) = 7 dB
Therefore, $IL = TL - C = 27 - 7 = 20 \, dB$

Referring to Fig. 4.8, $l = 4$ m, $b = 3$ m, $h = 3$ m *and* $d = 1$ m.

Now, making use of Eq. (4.37) we get the measurement surface area, $S_m$.

$$S_m = 2(4+2)(3+1) + 2(3+2)(3+1) + (4+2)(3+2) = 118 \text{ m}^2$$

Area of the room surface is given by

$$S = 2(20 \times 15) + 2 \times 6(20+15) = 1020 \text{ m}^2$$
$$\bar{\alpha} = 0.1 \ (given)$$

Thus, making use of Eq. (4.4) the room constant (at 500 Hz) works out to be

$$R = \frac{1020 \times 0.1}{1 - 0.1} = 113.3 \text{ m}^2$$

Finally, use of Eq. (4.36) yields the average value of SPL on the measurement surface:

$$L_p = 120 - 20 + 10 \ \log\left\{\frac{1}{118} + \frac{4}{113.3}\right\} = 86.4 \text{ dB}$$

## 4.5  Acoustics of Barriers

Acoustic barriers are often used for audio privacy between adjacent cabins in an office layout, partial protection of the road-side housing colony from the traffic noise, etc. Trees and bushes are also planted sometimes for landscaping as well as environmental noise control.

Sound diffracts around a finite barrier from all three sides. Low-frequency (or large-wavelength) waves bend around more efficiently than the high-frequency (or small-wavelength) waves. The effectiveness of barriers increases with Fresnel number, $N_i$, defined by [1]

$$N_i \equiv 2\delta_i / \lambda \tag{4.38}$$

where $\delta_i$ = difference in the diffracted path and the direct path between the source and the receiver (m).

$\lambda$ = wavelength, $c_0/f$ (m).

SPL at direct distance $r$ across the barrier in a room may be evaluated by incorporating $D$, the diffraction directivity factor, into Eq. (4.1). Thus, in the frequency band of interest, we can write [1].

$$L_p(r) = L_w + 10 \log\left(\frac{Q_B}{4\pi r^2} + \frac{4}{R}\right) \text{ dB} \tag{4.39}$$

where $Q_B$, the barrier directivity factor is product of the location-cum-inherent directivity factor $Q$ and the diffraction directivity factor $D$. Thus,

$$Q_B = Q \cdot D \tag{4.40}$$

where

$$D = \sum_{i=1}^{3} \frac{1}{3+10N_i} = \sum_{i=1}^{3} \frac{\lambda}{3\lambda + 20\delta_i} \tag{4.41}$$

If the barrier is extended to the ceiling or one of the walls, then that path is blocked and the summation in Eq. (4.41) must exclude it; the subscript '$i$' will no longer extend to 3. To be more precise, there would also be some contribution to $D$ from other secondary paths because of reflections from the ground, side walls and ceiling of the room, but these can be ignored from a practical point of view (for an accuracy of $\pm 1$ dB).

Comparing Eq. (4.39) with (4.1), insertion loss of a barrier within a room is given by

$$IL = 10 \log\left(\frac{\dfrac{Q}{4\pi r^2} + \dfrac{4}{R}}{\dfrac{Q_B}{4\pi r^2} + \dfrac{4}{R}}\right) \text{ dB} \tag{4.42}$$

where $Q_B$ is given by Eqs. (4.40) and (4.41).

It may be noted from Eq. (4.42) that for the barrier to be effective indoors, i.e., for a substantial value of IL, the reverberant field contribution 4/R must be much less than the direct field contribution $Q/(4\pi r^2)$. In other words, surface of the room must be acoustically treated. Alternatively, $r$ must be as small as feasible; i.e., the source as well as the receiver must be in the acoustical shadow of the barrier.

In a free field, however, the room constant R would tend to infinity, and then Eq. (4.42) would reduce to

$$IL(free\ field) = -10\ \log\ D \qquad (4.43)$$

Combination of Eqs. (4.41) and (4.43) yields

$$IL(free\ field) = -10\ \log\left[\lambda\left\{\frac{1}{3\lambda + 20\delta_1} + \frac{1}{3\lambda + 20\delta_2} + \frac{1}{3\lambda + 20\delta_3}\right\}\right]$$

$$(4.44)$$

For a long acoustic barrier (extending several wavelengths on either side) like a highway barrier, Eq. (4.44) reduces to

$$IL(long\ barrier,\ freefield) = -10\ \log\left[\frac{\lambda}{3\lambda + 20\delta_1}\right] \qquad (4.45)$$

For such a barrier, height is the only design parameter. Referring to Fig. 4.9,

$$\delta = (c+d) - (a+b) \qquad (4.46)$$

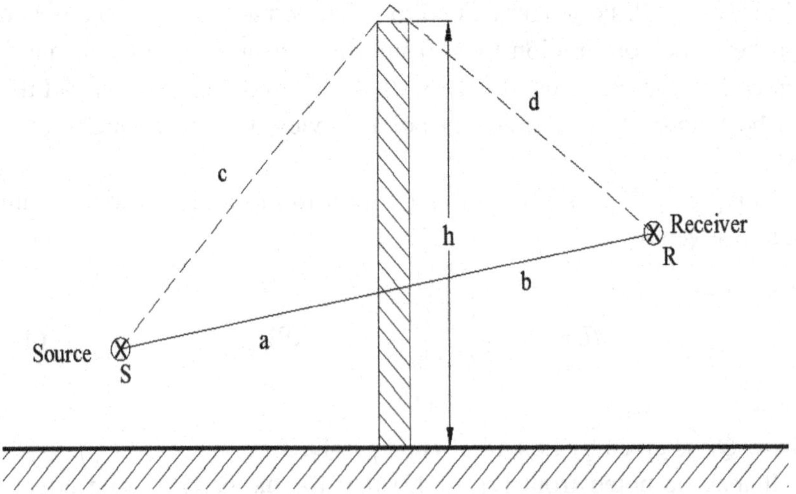

Fig. 4.9 Schematic of a long (semi-infinite) barrier of height $h$.

The trigonometric implications of Eq. (4.46) and Fig. (4.9) are as follows:

(a) The more the barrier height $h$, the more would be its insertion loss *IL*. As an important corollary, a highway barrier blocks tyre noise more effectively than the engine noise.
(b) Increasing the height beyond (say) 5 m is generally not a very cost-effective measure. One could explore the feasibility of building the acoustic barrier on the top of an intervening hill or earthen mound.
(c) The barrier should be located so that the source or receiver falls in its shadow, as it were. For example, a railway barrier should be located as near to the railway line or the railway colony as logistically feasible.
(d) When a highway or railway line has acoustic barriers on both sides, the barriers should be lined with an acoustically absorptive layer which can stand the elements (sun, rain, snow, etc.).

**Example 4.5** It is proposed to isolate the DG set of Example 4.4 from the rest of the shed by means of a wall-to-wall 5-m high barrier as shown in Fig. 4.9, instead of a stand-alone acoustic enclosure. Evaluate the insertion loss of the barrier in the 500-Hz octave band.

**Solution**

We make use of Eq. (4.42) along with Eqs. (4.40) and (4.41)
Wave length $\lambda = c_0 / f = 344 / 500 = 0.688$ m
As the DG set is on the floor, $Q = 2$

Assume the acoustic centre of the DG set to be at a height of 1 m; i.e.,
$$h_s = 1 \text{ m}$$

Let the receiver be also at about the same height, so that
$$h_R = 1 \text{ m}$$

Logistically, keeping in mind the dimensions of the DG set, let
$$a = b = 3 \text{ m}$$

Then, referring to Fig. 4.9, we have

$$c = d = \left\{ a^2 + \left( h - h_s \right)^2 \right\}^{1/2}$$

$$= \left\{ 3^2 + \left( 5 - 1 \right)^2 \right\}^{1/2}$$

$$= 5 \text{ m}$$

As per Eq. (4.46),

$$\delta = (5+5) - (3+3) = 4 \text{ m}$$

As the barrier stretches from wall to wall, diffraction will take place from the top only. Thus, use of Eq. (4.41) yields

$$D = \frac{\lambda}{3\lambda + 20\delta} = \frac{0.688}{3 \times 0.688 + 20 \times 4} = 0.0084$$

Distance of the receiver from the source, $r = a + b = 3 + 3 = 6$ m
From Example 4.4, room constant (at 500 Hz), R = 113.3 m$^2$

Substituting those data in Eq. (4.42) yields

$$IL = 10 \, \log \left[ \frac{\dfrac{2}{4\pi \times 36} + \dfrac{4}{113.3}}{\dfrac{2 \times 0.0084}{4\pi \times 36} + \dfrac{4}{113.3}} \right] = 10 \, \log \left[ \frac{0.0044 + 0.0353}{0.00004 + 0.0353} \right]$$

$$= 0.5 \text{ dB only!}$$

Thus, the barrier is practically useless. This disappointing result indicates a very important design consideration: a barrier is ineffective in a reverberant room or shed. The room, particularly its ceiling, must be acoustically lined for the wall-to-wall barrier to offer substantial insertion loss. For example, if all available area of the four walls (above a height of 2 m) and ceiling of the room were lined with an acoustical material with $\alpha$ (*at* 500 Hz) = 0.94 (see Table 4.1 for rockwool blanket of 500 mm thickness), then using Eq. (4.5),

$$\bar{\alpha} = \frac{\{2(20+15)(6-2) + 20 \times 15\} \times 0.94 + 20 \times 15 \times 0.1}{2(20+15) \times 6 + 2 \times 20 \times 15} = 0.564$$

For this value of $\bar{\alpha}$, room constant R will be

$$R = \frac{1020 \times 0.564}{1 - 0.564} = 1319 \text{ m}^2$$

Then, $\dfrac{4}{R} = \dfrac{4}{1319} = 0.003$

And the new value of IL is given by

$$IL = 10 \ \log\left[\frac{0.0044 + 0.003}{0.00004 + 0.003}\right] = 3.9 \quad dB$$

This value is typical of acoustic barriers. It is substantially higher than the value of 0.5 dB of the original (unlined) room. However, it is much lower than *IL* of 20 dB offered by the acoustic enclosure of Example 4.4. Therefore, for indoor use in industrial sheds, acoustic barrier is not a design option; one must use a stand-alone acoustic enclosure or an acoustic hood instead.

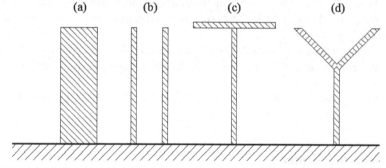

Fig. 4.10 Enhancing the insertion loss of a barrier: (a) thick barrier, (b) compound (double) barrier, (c) barrier with a flat cap, and (d) barrier with a forked top.

Other ways of increasing the *IL* of the barrier are shown in Fig. 4.10. As a corollary of the barrier shapes shown in this figure, any buildings, sheds, store rooms, etc., would yield high values of insertion loss. This feature is used extensively by knowledgeable architects while designing residential layouts along or near the busy highways and railway lines.

Equation (4.38) indicates that a barrier of given height will be more effective at higher frequencies than at lower frequencies. In fact, at very high frequencies, sound travels in a straight line like light rays, and therefore the barrier acts as a very effective reflector; there is little diffraction around the barrier.

## References

1. Irwin, J. D. and Graf, E. R., Industrial Noise and Vibration Control, Prentice-Hall, Englewood Cliffs, NJ, USA, (1979).
2. Sabine, W. C., Collected Papers on Acoustics, American Institute of Physics, New York, (1993).
3. Bies, D. A., Uses of anechoic and reverberant rooms, Noise Control Engineering Journal, 7, pp. 154-163, (1976).
4. International Standards Organization, Acoustics: Measurement of sound absorption in a reverberation room, ISO 354, (1985).
5. American Society of Testing Materials, Standard test method for sound absorption coefficient by the reverberation room method, ASTM C423-99a, (1984).
6. Bies, A. and Hansen, C. H., Engineering Noise Control: Theory and Practice, Fourth Edition, Spon Press, London, (2009).
7. Harris, C. M., Noise Control in Buildings, McGraw Hill, New York, (1994) (Appendix 3: Tables of Sound Absorption Coefficients, compiled by C.M. Harris and Ron Moulder).
8. American National Standard 'Engineering methods for the determination of sound power levels of noise sources in a special reverberation test room' ANSI S12.33-1990 (also, ISO 3742).
9. Crocker, M. J., Handbook of Noise and Vibration Control, (Ed.), Chapter 56, John Wiley, New York, (2007).

## Problems in Chapter 4

**Problem 4.1** Space-averaged sound pressure level in a $8m \times 5m \times 4m$ (high) reverberant room with the reverberation time of 2 seconds in the 1000 Hz octave frequency band, produced by a loudspeaker source of 10 *W* electrical power, is 100 dB. What is the acoustical to electrical power conversion efficiency of the loudspeaker in the particular frequency band?

**[Ans.: 0.346%]**

**Problem 4.2**   Reverberation time in the 500 Hz octave band of a 7m×5m×4m (high) room is 2 s when bare, and 1.3 s when a 3m×3m area of its floor is covered with an acoustical blanket under test. Evaluate the acoustical power absorption coefficient of the test blanket.

**[Ans.: 0.67]**

**Problem 4.3**   Average SPL due to a window air conditioner in a bare (unfurnished) 5m×4m×3m (high) room with $\alpha = 0.05$ is 60 dB in the 500-Hz octave band. If the entire ceiling of the room were covered with a 13 *mm* thick mineral tile (directly fixed to the ceiling slab), what would be the reduced value of *SPL* in the same frequency band?

**[Ans.: 53.6 *dB*]**

**Problem 4.4**   A portable genset, located in a corner of the floor of 8*m*×5*m*×3*m* (high)  room with average sound absorption coefficient of 0.1 in the 250-Hz octave band, radiates sound power level of 86 dB in that band.   Estimate SPL in the middle of the room. At what distance, will the direct sound field be equal to the diffuse field?

**[Ans.: 80 *dB*; 1.67 m]**

**Problem 4.5**   Making use of the Mass Law, evaluate TL for a 10 mm thick aluminium panel over all the eight major octave frequency bands (63 Hz to 8000 Hz), when the panel is
   (a)  immersed in water
   (b)  in air at 25°C
   **[Ans.: Water :  0.0,   0.0,   0.1,   0.3,   1.2,   3.7,   8.0, 13.4 dB**
   **Air : 42.2, 48.2, 54.3, 60.3, 66.3, 72.3, 78.3, 84.4 dB]**

**Problem 4.6**   What should be the minimum thickness of the impervious steel plate of an acoustic enclosure, lined on the inside with 100 mm thick, 64 kg/m$^3$ density mineral wool so that the enclosure has an insertion loss of at least 25 dB at 500 Hz.

**[Ans.: 2.1 mm]**

**Problem 4.7**   An out-door compressor is generating a sound power level (SWL) of 120 dB at 500 Hz. The Government requires sound pressure level (SPL) at the property line 10 m away not to exceed 75 dB during the day and 70 dB during the night operation in an industrial area. Evaluate the minimum insertion loss for which the acoustic enclosure for the compressor should be designed in order to meet the statutory limit for 24-hour operation of the compressor.

**[Ans.: 22 dB]**

**Problem 4.8** A railway line is provided with a sufficiently long, 5m high acoustic barrier in order to give some acoustical protection to a parallel row of residential houses 50 m away. The barrier is located 5 m away from the centerline of the railway tracks. Assuming the source to be at a height of 1 m from the ground, what would be the noise reduction at 250 Hz for a resident at a height of

 (a)  2 m (ground floor), and
 (b)  6 m (first floor)?

**[Ans.: (a) 13.9 dB, (b) 13.0 dB]**

**Problem 4.9** A 4m high and 5m long acoustic barrier is erected in the centre of a 10 m×8 m×5 m (high) room with average sound power absorption coefficient of 0.2 in the 500-Hz octave band, as shown in the figure below. What is the insertion loss of the barrier in this same frequency band for a set of source (S) and receiver (R) if both are located 1.5 m from the ground and 3 m on opposite sides of the barrier? What would be the IL of the same barrier in free field?

**[Ans.: 0.27 dB; 12.2 dB]**

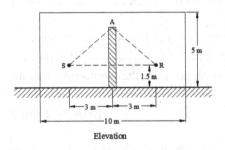

Elevation

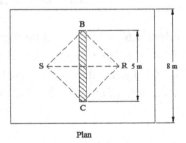

Plan

# Chapter 5

# Mufflers and Silencers

The terms 'muffler' and 'silencer' are often used interchangeably. They are used extensively on the intake as well as exhaust systems of the reciprocating internal combustion engines, compressors, fans, blowers, gas turbines, the heating ventilation and air-conditioning (HVAC) systems, high-pressure vents and safety valves. Practically, the intake and/or exhaust (or discharge) systems of all flow machinery are fitted with mufflers. Automotive engines are invariably provided with exhaust mufflers. Therefore, the theory and practice of exhaust mufflers has history of over a hundred years.

Passive mufflers are of two types: the reactive (or reflective) type and absorptive (or dissipative) type. Reactive mufflers work on the principle of impedance mismatch. Making use of sudden changes in the area of cross-section, perforated elements, resonators, etc, the incoming or incident energy is reflected back to the source. In fact, a combination of such elements helps to reduce the acoustic load resistance faced by the source to much less than the characteristic impedance of the exhaust (or intake) pipe so that the source produces much less noise than it would produce into an anechoic load (or termination). An absorptive muffler, on the other hand, does not alter the sound produced by the source, but converts it into heat as sound propagates through its absorptive passages – acoustically absorptive linings.

Active noise control in a duct consists in making use of a secondary source of noise and an adaptive digital control system in order to produce nearly zero impedance at the junction. Thus, an active noise control system produces an acoustical short-circuit effectively muffling the primary as well as secondary sources of sound. Active noise control system is most effective at low frequencies (50 to 500 Hz) whereas passive mufflers work best in the middle and high frequencies. The reactive mufflers, however, can be specially configured for low frequency attenuation, as will become clear later in the chapter.

Design or selection of a muffler is based on the following considerations:

(i)    Adequate insertion loss so that the exhaust (or intake) noise is reduced to the level of the noise from other components of the engine (or compressor or fan, as the case may be), or as required by the environmental noise pollution limits;
(ii)   Minimal (or optimal) mean pressure drop so that the source machine does not have to work against undue or excessive back pressure. (This is particularly applicable to fans or blowers which would stall under excessive back pressure);
(iii)  Size restrictions, particularly under the vehicle;
(iv)   Weight restrictions;
(v)    Durability, particularly in view of sharp thermal gradients in rain or on wet roads;
(vi)   Cost effectiveness is often the most important design criterion.

## 5.1  Electro-Acoustic Modeling

We advocate here use of the direct electro-acoustic analogies, where acoustical pressure and mass (or volume) velocity correspond to the electromotive force (or voltage) and current, respectively. Thus, acoustical impedance would correspond to the electrical impedance, and we can make use of the electrical analogous circuit for representing one-dimensional acoustical filters or mufflers as shown in Fig. 5.1.

As per Thevenin theorem, working in the frequency domain, any linear time-invariant source may be presented by an open-circuit voltage or its acoustical analogue $p_s$ and internal impedance $Z_s$, inline with the source as shown in Fig. 5.1(a). This is called the voltage (or pressure) representation of the source. Alternatively, as per Norton theorem, the sources can be represented by a current, or its acoustical analogue, mass velocity $v_s$, with source impedance $Z_s$ in the shunt position, as shown in Fig. 5.1(b). It can be easily shown [1] that the two representations are equivalent – they deliver the same current or velocity against a given load impedance – provided $v_s = p_s/Z_s$ .

Acoustic load impedance faced by the source is given by

$$\zeta_n = p_n/v_n \tag{5.1}$$

When $Z_s \to 0$, then Fig. 5.1(a) shows that $p_n = p_s$ for all values of the load impedance $\zeta_n$, and the source acts as a constant pressure source. When $Z_s \to \infty$, then Fig. 5.1(b) shows that $v_n = v_s$ for all values of the load impedance $\zeta_n$, and the source acts a constant velocity source. Thus, a zero impedance source is a constant pressure source and an infinite impedance source is a constant velocity source.

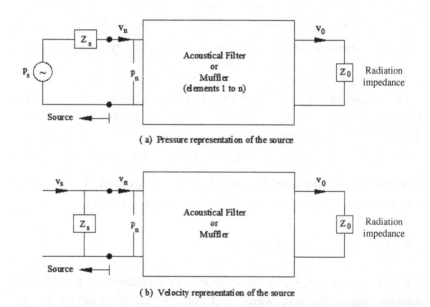

( a) Pressure representation of the source

( b) Velocity representation of the source

Fig. 5.1 Electrical analogous circuit of the muffler, radiation impedance and two equivalent representations of the source.

For use on an engine, the muffler consists of an exhaust pipe, the muffler proper and tail pipe, as shown in Fig. 5.2.

In the absence of any muffler, the source would face the radiation impedance $Z_0$ instead of $\zeta_n$. For typical applications (engines), effect of the mean temperature and density gradients is relatively small, and the exhaust pipe diameter is generally equal to the tail pipe diameter, and therefore, $Y_n = Y_1$. Referring to Fig. 5.1(b), it can be shown that insertion loss of a given muffler is given by [1]

$$IL = 20 \, \log \left| \frac{Z_s}{Z_s + Z_0} \frac{v_s}{v_0} \right| \qquad (5.2)$$

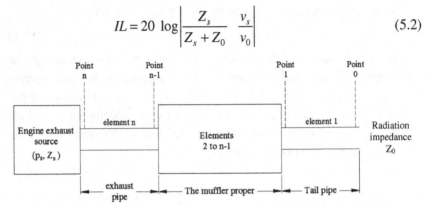

Fig. 5.2 Nomenclature of acoustical elements and points for a typical engine exhaust system.

It can also be shown [1] that transmission loss is the limiting value of $IL$ for anechoic source $(Z_s = Y_n)$, anechoic load $(Z_0 = Y_1)$, and $Y_n = Y_1 (= Y_0, \, say)$. Symbolically,

$$TL = Lim \; IL \; as \; Z_s = Z_0 \to Y_0 \qquad (5.3)$$

where $Y_0$ is the characteristic impedance of the exhaust pipe as well as the tail pipe (the two are assumed here to be of the same cross-section).

## 5.2   Transfer Matrix Modeling

The velocity ratio $v_s/v_0$ in Eq. (5.2) may be calculated by means of the transfer matrix method, which is ideally suited for analysis of one-dimensional systems like acoustical filters (or mufflers), vibration isolators, electrical wave filters, etc. Making use of the matrizant approach along with the basic governing equations, namely the mass continuity equation, momentum equation, isentropicity relation, and working in the frequency domain (that is, assuming harmonic time dependence $e^{j\omega t}$ for both the state variables, $p$ and $v$ at all points of the muffler), one can derive transfer matrices for a variety of elements that constitute the present-day automotive mufflers [1, 2]. Some of these basic elements are shown in Figs. 5.3 to 5.36. In each of these figures, points u and d denote 'upstream' and 'downstream', respectively. Transfer matrices connect state variables $p$ and $v$ at point u and those at point d.

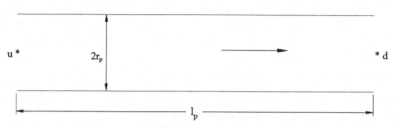

Fig. 5.3 Uniform diameter, rigid wall tube/duct/pipe.

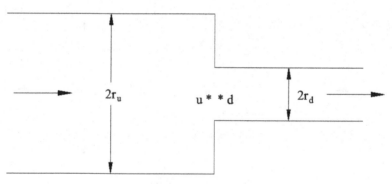

Fig. 5.4 Sudden contraction.

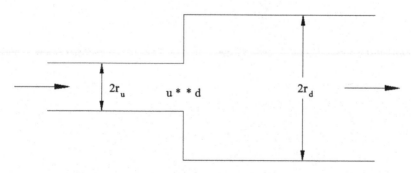

Fig. 5.5 Sudden expansion.

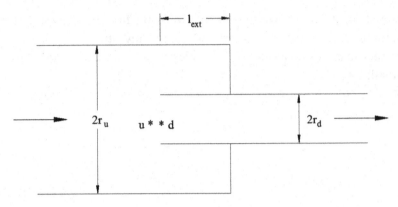

Fig. 5.6 Extended outlet.

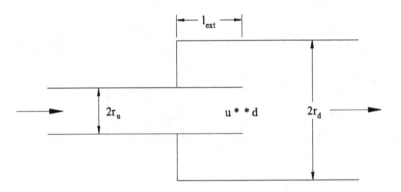

Fig. 5.7 Extended inlet.

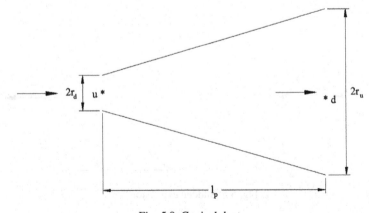

Fig. 5.8 Conical duct.

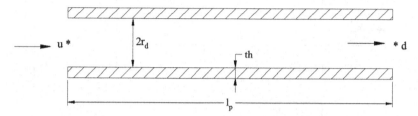

Fig. 5.9 Hose (or a uniform area tube with compliant wall).

Fig. 5.10 Concentric tube resonator.

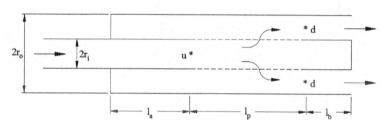

Fig. 5.11 Cross-flow expansion.

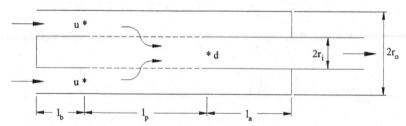

Fig. 5.12 Cross-flow contraction.

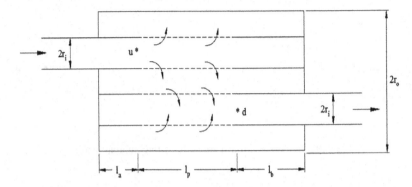

Fig. 5.13  Cross-flow, three-duct, closed-end element.

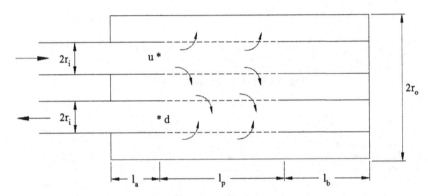

Fig. 5.14  Reverse-flow, three-duct, closed-end element.

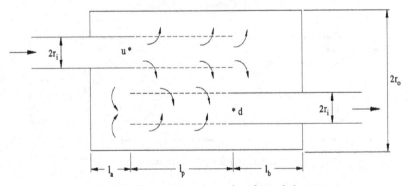

Fig. 5.15  Three-duct, open-end perforated element.

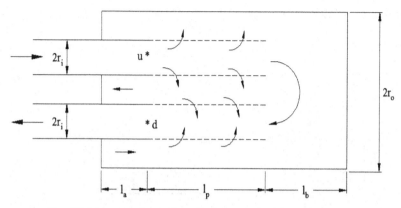

Fig. 5.16  Reverse-flow, three-duct open-end perforated element.

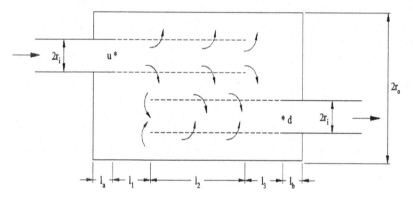

Fig. 5.17  Extended (non-overlapping) perforation cross-flow, open-end element.

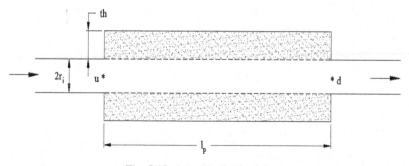

Fig. 5.18  Acoustically lined duct.

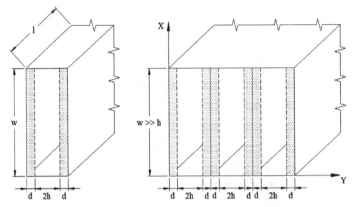

Fig. 5.19 Parallel baffle muffler and a constituent rectangular duct lined on two sides.

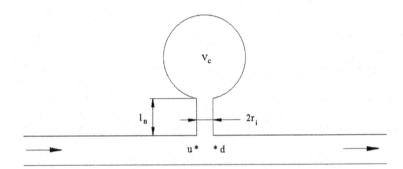

Fig. 5.20 Helmholtz resonator.

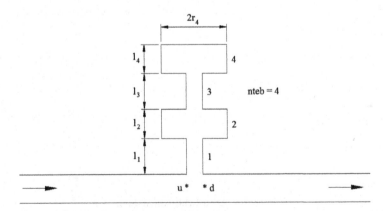

Fig. 5.21 Branch sub-system.

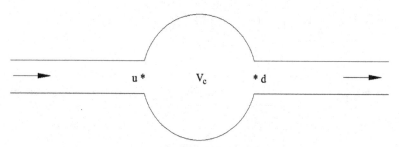

Fig. 5.22 Inline cavity.

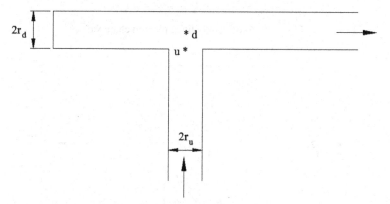

Fig. 5.23 Side inlet.

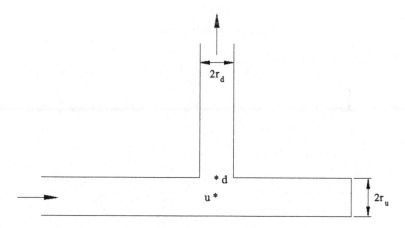

Fig. 5.24 Side outlet.

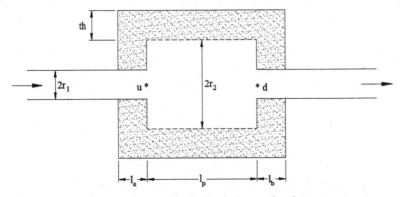

Fig. 5.25 Acoustically lined plenum chamber.

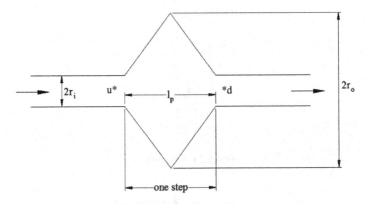

Fig. 5.26 Compliant bellows.

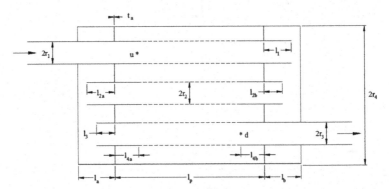

Fig. 5.27 Extended-tube 3-pass perforated element chamber.

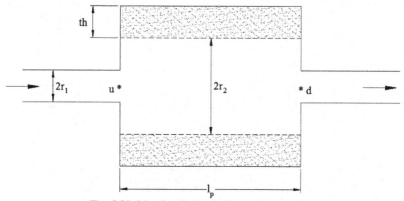

Fig. 5.28 Lined wall simple expansion chamber.

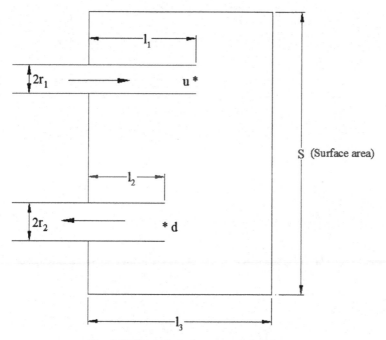

Fig. 5.29 Extended tube reversal chamber.

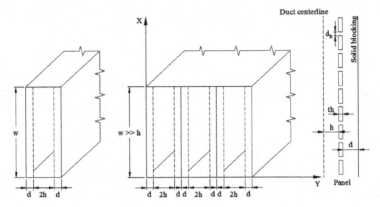

Fig. 5.30  Micro-perforated Helmholtz panel parallel baffle muffler and a constituent rectangular duct.

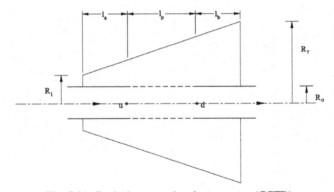

Fig. 5.31  Conical concentric tube resonator (CCTR).

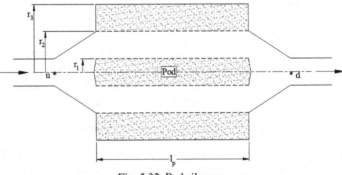

Fig. 5.32  Pod silencer.

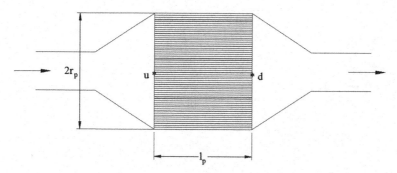

Fig. 5.33 Catalytic converter (capillary-tube monolith).

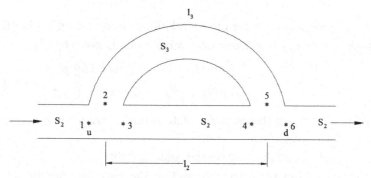

Fig. 5.34 Quincke tube.

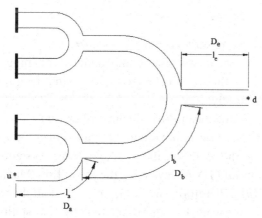

Fig. 5.35 Equal-length 4-cylinder runner manifold.

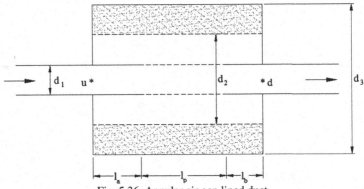

Fig. 5.36  Annular air gap lined duct.

For example, the transfer matrix relationship for a rigid wall uniform area pipe or tube (a one-dimensional waveguide) is given by [1]

$$\begin{bmatrix} p_u \\ v_u \end{bmatrix} = e^{-jMk_cl} \begin{bmatrix} \cos(k_c l) & jY_0 \sin(k_c l) \\ j\sin(k_c l)/Y_0 & \cos(k_c l) \end{bmatrix} \begin{bmatrix} p_d \\ v_d \end{bmatrix} \tag{5.4}$$

where $p_u$ and $v_u$ are the upstream state variables (acoustic pressure and mass velocity),

$p_d$ and $v_d$ are the downstream state variables,

The $2 \times 2$ square matrix connecting the two state vectors is the transfer matrix of the uniform pipe with moving medium,

$k_c = \dfrac{k_0}{1 - M^2}$ is the convective wave number,

$k_0 = \omega/c_0$ is the stationary wave number,

$Y_0 = c_0/S$ is the characteristic impedance of the pipe,

$S$ and $l$ are the area of cross-section and length of the pipe, respectively,

$M = U/c_0$ is the mean flow Mach number, and

$U$ is the mean flow axial velocity averaged over the cross-section.

Typically, in the case of the intake/exhaust systems of engines, compressors, fans and HVAC systems, the mean flow Mach number $M$ is of the order of 0.1. Therefore, $M^2 \ll 1$, $k_c \approx k_o$, and the transfer matrix in Eq. (5.4) may well be approximated by its stationary-medium counterpart:

$$\begin{bmatrix} \cos(k_0 l) & jY_o \sin(k_0 l) \\ \dfrac{j\sin(k_0 l)}{Y_0} & \cos(k_0 l) \end{bmatrix} \tag{5.5}$$

In fact, the convective effect of mean flow, by virtue of which the forward wave moves faster $(c_0 + U)$ and the reflected wave moves slower $(c_0 - U)$, may be neglected in the whole of muffler analysis as a first approximation. However, the dissipative effect of mean flow at the area discontinuities and perforates, because of flow separation, plays a crucial role in the muffler performance and therefore must be incorporated appropriately.

A muffler can often be conceptualized as a cascade of some of the basic elements shown in Figs. 5.3 to 5.36. This is illustrated in Fig. 5.37. Comparison of the muffler of Fig. 5.37 with basic elements shown in Figs. 5.3 – 5.7 indicates that elements 1, 3, 5, 7 and 9 are uniform tubes, element 2 is a sudden contraction, element 8 is a sudden expansion, element 4 is an extended inlet and element 6 is an extended outlet. The overall transfer matrix of the muffler proper can then be constructed by successive multiplication of the transfer matrices of its constituent elements. Let this overall transfer matrix be denoted as

$$[T] \equiv \begin{bmatrix} T_{11} & T_{12} \\ T_{21} & T_{22} \end{bmatrix} \tag{5.6}$$

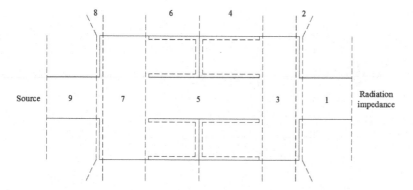

Fig. 5.37 Decomposition of a muffler into basic constituent elements.

Then, it can be shown that *TL* is given by the expression [1]:

$$TL = 10 \ \log\left[\left\{\frac{1+M_u}{1+M_d}\right\}^2 \frac{Y_d}{Y_u} \left|\frac{T_{11}+T_{12}/Y_d+T_{21}.Y_u+T_{22}Y_u/Y_d}{2}\right|^2\right] \quad (5.7)$$

When $Y_d = Y_u$ (or, $S_d = S_u$), then Eq. (5.7) simplifies to

$$TL = 20 \ \log\left|\frac{T_{11}+T_{12}/Y_0+T_{21}.\ Y_0+T_{22}}{2}\right| \quad (5.8)$$

where $Y_0 = Y_d = Y_u$. It may be recalled that subscripts $u$ and $d$ denote upstream and downstream, respectively.

This expression applies within the assumptions and simplifications described above, and is often sufficient for synthesizing a preliminary muffler configuration [1, 3-5].

## 5.3    Simple Expansion Chamber (SEC)

The simplest muffler configuration is a Simple Expansion Chamber (SEC) shown in Fig. 5.38. It was first conceptualized in the lumped-element model as a compliance (capacitance) sandwiched between two inertances (inductances), analogous to a low-pass filter in the electrical network theory. For stationary medium, making use of the distributed-element transfer matrix of Eq. (5.5) for the chamber of length *l*, and combining it with the *TL* expression (5.7) yields the classical relationship [6]

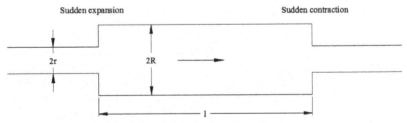

Fig. 5.38  A simple expansion chamber.

$$TL = 10 \ \log\left[1+\left\{\frac{m-1/m}{2}\sin\left(k_0 l\right)\right\}^2\right] \quad (5.9)$$

where $m$ is the area expansion ratio, $m = (R/r)^2$, and the product $k_0 l$ is called Helmholtz number or non-dimensional frequency.

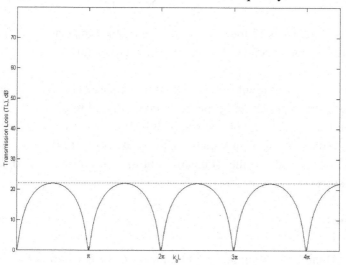

Fig. 5.39 TL of a simple expansion chamber.

Equation (5.9) is plotted in Fig. 5.39 for $m = 25$. The following features are noteworthy:

(i) The TL curve of a simple expansion chamber consists of periodic domes, with sharp troughs occurring at integral multiples of $\pi$, and peaks occurring at odd multiples of $\pi/2$.

(ii) Peak values of $TL$ is approximately $20 \log(m/2)$ for $m \gg 1$. Thus, the larger the expansion ratio, the higher the peaks (or domes).

(iii) As $m \to 1$, $TL \to 0\ dB$. This shows that $TL$ of a uniform pipe is zero.

(iv) Observations (ii) and (iii), when read together, indicate that in reactive mufflers, a sudden change in area of cross-section $S$ results in a sudden change or jump in characteristic impedance $Y_0 (= c_0 / S)$. In fact, making use of Eq. (5.7) it can be shown [1] that in stationary medium for sudden expansion (Fig. 5.5) and sudden contraction (Fig. 5.4), both characterized by a unity transfer matrix (representing equality of pressure and mass velocity across the junction) $TL$ is given by the common expression:

$$TL = 10 \ \log\left[\frac{\left(S_u + S_d\right)^2}{4 \ S_u S_d}\right] \qquad (5.10)$$

This is the basis of the concept of 'impedance mismatch' which is a fundamental principle of reactive or reflective mufflers.

**Example 5.1** For a simple expansion chamber shown in Fig. 5.38, radius of the exhaust pipe and tail pipe is 20 mm, and radius and length of the expansion chamber are 60 mm and 300 mm, respectively. Find the lowest frequency range in which its TL is at least 10 dB. Assume the medium to be air at the standard atmospheric pressure and $25^0$ C temperature.

**Solution**

$r = 20$ mm, $R = 60$ mm, $l = 500$ mm, $c_0$ at $25^0 C = 346$ m / s,

and $TL \geq 10$dB.

Area expansion ratio, $m = \left(R / r\right)^2 = \left(60 / 20\right)^2 = 9$.

Referring to Eq. (5.9) we have

$$10 = 10 \ \log\left[1 + \left\{\frac{9 - 1/9}{2}\sin\left(k_0 l\right)\right\}^2\right]$$

This gives, $\sin\left(k_0 l\right) = 0.675$

Thus, referring to the first or lowest frequency dome of Fig. 5.39,

$$k_0 l = 0.741 \quad \text{and} \quad \pi - 0.741 \ \text{radians}$$

or

$$\frac{2\pi f}{346} \times 0.3 = 0.741 \quad \text{and} \quad 2.4$$

or

$$f = 136 \ \text{Hz} \quad \text{and} \quad 440.5 \ \text{Hz}$$

Hence, for the specified simple expansion chamber, *TL* would be 10 dB or more in the frequency range 136 Hz to 440.5 Hz.

## 5.4 Extended-Tube Expansion Chamber (ETEC)

Schematic of a typical extended-tube expansion chamber (ETEC) is shown in Fig. 5.40. It indicates that it consists of extended inlet tube and extended outlet tube of extension lengths $l_a$ and $l_b$, respectively. An extended tube acts as a branch or shunt impedance in an electrical analogous circuit where the mass velocity (or current) gets divided. The two branch impedances in Fig. 5.40 are given by [1]

$$Z_a = -jY_a \cot(kl_a) \quad \text{and} \quad Z_b = -jY_b \cot(kl_b) \tag{5.11}$$

where $k$ is wave number, $k = \omega / c_0$; $Y_a = Y_b = c_0 / S_{ann}$ is the characteristic impedance of the annulus; and $S_{ann}$ is the area of cross-section of the annulus.

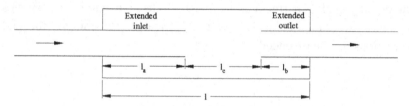

Fig. 5.40 An extended-tube expansion chamber.

Transmission loss of the ETEC of Fig. 5.40 would tend to infinity (i.e., no sound would be transmitted downstream) if $Z_a$ or $Z_b$ tended to zero (analogous to a short-circuit in the electrical network). This would happen when

$$kl_a = (2m-1)\pi/2 \quad \text{or} \quad kl_b = (2n-1)\pi/2 \tag{5.12}$$

where $m$ and $n$ are integers ($m$, $n$ = 1,2,3,............).

This interesting property can be utilized to cancel some of the troughs in the *TL* curve (Fig. 5.39) of the corresponding simple expansion chamber (Fig. 5.38) of length $l$. It can be shown that [4, 7]

if $l_a = l/2$, then troughs 1, 3, 5, 7, .... would be nullified/tuned out, and

if $l_b = l/4$, then troughs 2, 6, 10, 14, .... would be nullified/tuned out.

$$\tag{5.13}$$

This is shown in Fig. 5.41. It may be noted that all troughs except 4, 8, 12, ……. are tuned out. Moreover, there is an overall lifting of the peak value of *TL*. This is the basic principle of a double-tuned extended-tube expansion chamber [4, 7].

In actual practice, however, relations (5.12) and (5.13) do not hold because they do not take into account the effect of three-dimensional (3-D) evanescent waves generated at the area discontinuities. This effect manifests itself as end corrections such that [8, 9]

acoustic length = geometrical (or physical) length + end correction

The end corrections of sudden expansion and sudden contraction have been calculated in the literature analytically [8] and numerically [9]. The numerical methods make use of FEM or BEM techniques. Similarly, for the extended inlet and outlet, the end correction has been evaluated numerically and experimentally [7, 10]. Extensive parametric studies [10] have resulted in the following empirical expression for the end corrections for the extended inlet/outlet:

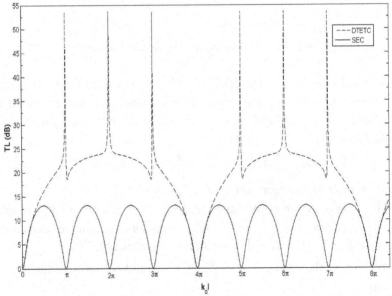

Fig. 5.41 Comparison of TL of a double-tuned expansion chamber with that of a simple expansion chamber.

$$\frac{\delta}{d} = a_0 + a_1 \left(\frac{D}{d}\right) + a_2 \left(\frac{t_w}{d}\right) + a_3 \left(\frac{D}{d}\right)^2 + a_4 \left(\frac{D}{d}\frac{t_w}{d}\right) + a_5 \left(\frac{t_w}{d}\right)^2 \quad (5.14)$$

Here,
$a_0 = 0.005177$, $a_1 = 0.0909$, $a_2 = 0.537$, $a_3 = -0.008594$, $a_4 = 0.2616$, $a_5 = -5.425$; $d$ and $t_w$ are diameter and wall thickness of the inner inlet/outlet tube; and $D$ is the (equivalent) shell diameter.

It was noted that the end correction for a given extended length is the same whether it is used for extended inlet or for extended outlet [10].
Relations (5.13) are modified now as follows:

$$l_{g,a} = \frac{l}{2} - \delta \quad \text{and} \quad l_{g,b} = \frac{l}{4} - \delta \quad (5.15)$$

Predictions of the 1-*D* model match with those of the 3-*D* model only when we add the end corrections to the geometric lengths for use in the 1-*D* model, as shown in Fig. 5.41.

Practically, double-tuning aims at tuning out the first three troughs in the TL curve of the corresponding simple expansion chamber. This is because the unmuffled noise of engines, compressors, fans, etc. predominates at lower frequencies (upto 1000 Hz) and the higher-order modes get cut on (start propagating) at higher frequencies.

**Example 5.2** Re-design the simple expansion chamber of Example 5.1 as a double-tuned extended-tube expansion chamber. Assume that the wall thickness of the inner tube is 1.25 mm. Calculate the lowest range of frequency over which TL would exceed 10 dB.

**Solution**

Referring to the extended-tube expansion chamber of Fig. 5.40, we make use of Eqs. (5.14) and (5.15) to determine the required extensions $l_{g,a}$ and $l_{g,b}$ as follows.

$d = 2r = 40$ mm, $D = 2R = 120$ mm, $D/d = 120/40 = 3$, $l = 300$ mm, $t_w = 1.25$ mm

Substituting these dimensions in Eq. (5.14) yields

$$\frac{\delta}{40} = 0.005177 + 0.0909 \times 3 + 0.537\left(\frac{1.25}{40}\right) - 0.008594(3)^2$$

$$+ 0.02616\left(\frac{3 \times 1.25}{40}\right) - 5.425\left(\frac{1.25}{40}\right)^2$$

$$= 0.2145$$

Thus, the end correction $\delta = 40 \times 0.2145 = 8.6$ mm.

Substituting the values of chamber length $l$ and end correction $\delta$ in Eq. (5.15) gives the required physical or geometric lengths of the extended inlet, $l_{g,a}$ and extended outlet, $l_{g,b}$:

$$l_{g,a} = \frac{300}{2} - 8.6 = 141.4 \text{ mm}$$

$$l_{g,b} = \frac{300}{4} - 8.6 = 66.4 \text{ mm}$$

Extending the frequency-range logic of Example 5.1 to Fig. 5.41 for the double-tuned extended tube expansion chamber muffler, we find that $TL$ would exceed 10 dB over the range

$$k_0 l = 0.741 \text{ to } 4\pi - 0.741 \quad \text{radians}$$

or

$$\frac{2\pi f}{346} \times 0.3 = 0.741 \text{ to } 11.825 \quad \text{radians}$$

or

$$f = 136 \text{ Hz} \text{ to } 2171 \text{ Hz}$$

Comparing this wide range with that of the corresponding simple expansion chamber of Example 5.1 (136 Hz to 440.5 Hz), reveals the tremendous advantage of double tuning, as may be noted from Fig. 5.41.

## 5.5 Extended Concentric Tube Resonator (ECTR)

Schematic of a concentric tube resonator (CTR) is shown in Fig. 5.10. When combined with the ETEC of Fig. 5.40, it results in an extended concentric tube resonator (ECTR) as shown in Fig. 5.42.

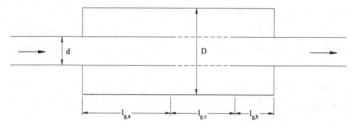

Fig. 5.42 Schematic of an extended concentric tube resonator (ECTR).

Following the control volume approach of Sullivan and Crocker [10], analysis of a concentric tube resonator involves modeling of the convected plane wave propagation in the inner perforated tube and the annular cylindrical cavity coupled thorough impedance of the perforation [1, Chap.3]. For large porosity, TL of a concentric tube resonator resembles the domed structure of the TL of the corresponding simple expansion chamber. Therefore, the extended concentric tube resonator of Fig. 5.42 may be looked upon as an extended-tube expansion chamber (ETEC) with a perforated bridge. This point of view is particularly exciting inasmuch as we can try to tune it like we did for the extended-tube expansion chamber.

Provision of a perforate bridge between the inlet and outlet of an extended-tube expansion chamber has advantages of little aerodynamic noise, minimal pressure drop and increased mechanical strength and durability. However, acoustic action of the resultant extended concentric tube resonator (ECTR) is very different from the corresponding double-tuned extended-tube chamber. Distributed hole-inertance of the CTR replaces the lumped inertance at the area discontinuities due to evanescent waves in the case of the corresponding extended-tube chambers.

The effective acoustic lengths are calculated precisely using 1-D analysis of the ECTR. The difference of these acoustical lengths and the quarter wave resonance lengths, i.e. half and quarter of chamber lengths are termed as differential lengths. There are many variables along with temperature dependence that affect the geometrical length required to tune the ECTR, so one needs to use 1-D analysis to estimate acoustical length and calculate the required physical lengths from the differential lengths and end corrections.

The differences between the two lengths (acoustical and geometric) are termed here as end corrections. These are the consequence of the inertance of perforates.

The following least-squares fit has been developed for the differential length normalized with respect to the inner-tube diameter [11]

$$\frac{\Delta}{d} = 0.6643 - 2.699\sigma + 4.522\sigma^2; \quad \Delta = \Delta_a = \Delta_b \qquad (5.16)$$

where $\sigma$ is porosity of the perforates (as a fraction). Eq. (5.16) is applicable for $\sigma$ ranging from 0.1 to 0.27.

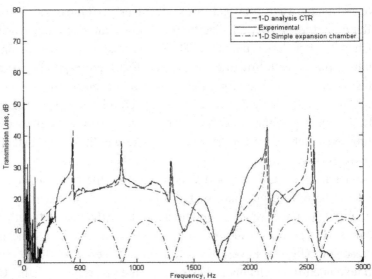

Fig. 5.43 Comparison between 1-D prediction and experimental measurements for ECTR of Fig. 5.42 with 19.6% porosity [11].

Differential lengths are calculated from Eq. (5.16) for the particular porosity and inner tube diameter and these are used to estimate the initial values of acoustical lengths $\left(l_{g,a} = l/2 - \Delta, \; l_{g,b} = l/4 - \Delta\right)$. With the help of the 1-D analysis we can increase/decrease these lengths such that the chamber length troughs are nullified effectively.

Predictions of the 1-D model match with those observed experimentally, and the end corrections for this particular case are almost zero, as shown in Fig. 5.43. In particular, the first three peaks of the 1-D curve match exactly with experimental results, as shown in Fig. 5.43.

Thus, one makes use of the 1-D analysis along with precise differential lengths and end corrections to tune the extended concentric tube resonators so as to lift or tune out three-fourths of all troughs that characterize the TL curve of the corresponding simple expansion chamber muffler. This makes the tuned ECTR a viable design option.

**Example 5.3** To reduce the mean pressure drop and the aerodynamic noise generation in the extended-tube expansion chamber muffler of Example 5.2, it is proposed to provide a perforate bridge with 19.6 % porosity so as to convert it into an extended concentric tube resonator, as shown in Fig. 5.42. Estimate the extended lengths $l_{g,a}$ and $l_{g,b}$.

**Solution**

Given: Chamber length, $l = 0.3$ m
Inner tube diameter, $d = 2 \times 0.02 = 0.04$ m
Porosity of the perforated section, $\sigma = 0.196$

The differential length $\Delta$ is given by Eq. (5.16). Thus,

$$\frac{\Delta}{0.04} = 0.6643 - 2.699 \times 0.196 + 4.522(0.196)^2$$

or

$$\Delta = 0.04 \times 0.31 = 0.0124 \text{ m} = 12.4 \text{ mm}$$

Now, first estimates for the extension lengths are given by

$$l_{g,a} = \frac{l}{2} - \Delta \quad and \quad l_{g,b} = \frac{l}{4} - \Delta$$

Thus,          $l_{g,a} = \dfrac{0.3}{2} - 0.0124 = 0.1376 \text{ m} = 137.6 \text{ mm}$

and

$l_{g,b} = \dfrac{0.3}{2} - 0.0124 = 0.0626 \text{ m} = 62.6 \text{ mm}$

Finally, length of the perforate (Fig. 5.42) is given by

$$l_{g,c} = l - \left( l_{g,a} + l_{g,b} \right)$$
$$= 300 - (137.6 + 62.6) = 99.8 \text{ mm}$$

Starting with these estimates, we need to do a rigorous analysis of the extended-tube *CTR* making use of the theory outlined in Section 3.8 of Ref. [1] incorporating mean flow, in order to fine-tune $l_{g,a}$ and $l_{g,b}$ for obtaining the *TL* curve shown in Fig. 5.43.

## 5.6  Plug Muffler

Schematic of a typical plug muffler is shown in Fig. 5.44. As the flow pattern indicates, the plug forces the mean flow to move out through the perforation into the annulus and then move back through the perforation to the inner tube. Thus, a plug muffler is a combination of a cross-flow expansion element (Fig. 5.11) and a cross-flow contraction element (Fig. 5.12) with a small uniform annular tube (Fig. 5.3) in between. Cross flow elements act as in-line lumped flow-acoustic resistance. Therefore, their acoustic performance (*TL*) as well as mean pressure drop increase as the open area ratio (OAR) decreases and the mean flow Mach number increases while remaining incompressible.

This important parameter is defined as

$$OAR \equiv \frac{flow \ area \ of \ the \ perforated \ section}{cross-sectional \ area \ of \ the \ perforated \ tube}$$

$$= \frac{\pi d l_p \sigma}{\dfrac{\pi}{4} d^2} = \frac{4 \sigma l_p}{d} \tag{5.17}$$

where $l_p$ and $\sigma$ are length and porosity of the perforated section (generally the same on both sides of the plug – see Fig. 5.44).

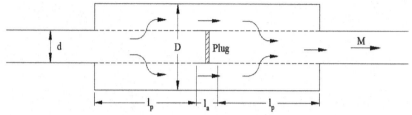

Fig. 5.44 Schematic of a plug muffler.

Figure 5.45 shows TL spectra of the plug muffler of Fig. 5.44 with OAR as a parameter. For $l_p = 20$ cm, $d = 4$ cm, as per Eq. (5.17), porosity works out to be $\sigma = 0.05$ OAR.

It is obvious from Fig. 5.45 that TL increases remarkably as OAR decreases. But then, as we shall see later in the section on back pressure, mean pressure drop increases drastically as OAR decreases. This calls for a compromise between back pressure and *TL*.

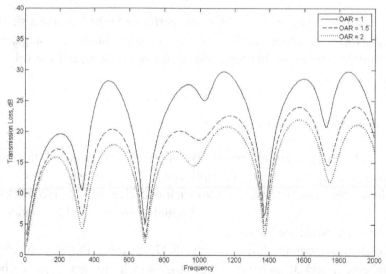

Fig. 5.45 TL of the plug muffler of Fig. 5.44 as a function of the open-area ratio, OAR.
$(d = 4$ cm, $D = 12$ cm, $l_p = 20$ cm, $l_a = 10$ cm, $M = 0.15)$.

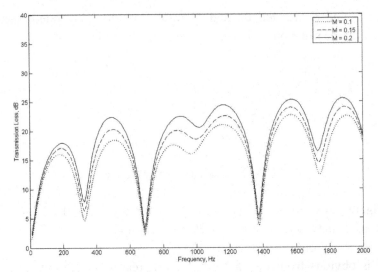

Fig. 5.46  Effect of the mean flow mach number $M$ on $TL$ of a plug muffler.
($d$ = 4 cm, $D$ = 12 cm, $l_p$ = 20 cm, $l_a$ = 10 cm, OAR = 0.15)

Mean flow Mach number, $M$, in the perforated tube has also a similar effect as may be observed from Fig. 5.46. Incidentally, the stagnation pressure drop increases in proportion to square of the mean flow velocity or Mach number.

## 5.7  Multiply Connected Muffler

Many of the present-day automotive mufflers are elliptical in cross-section and make use of multiply-connected elements. One such configuration would consist of two such end chambers connected by a uniform pipe, thereby making use of double flow reversal and inducing maximum acoustic interference.

Such a system is shown in Fig. 5.47, wherein the end-chambers (numbered 1 and 3) and connecting pipes which also act as pass tubes, are clearly shown. The lengths $L_c$ of the connecting pipes are much greater than the lengths of the elliptical end chambers $L_a$ and $L_b$. Rather than the time consuming FEM process, which involves geometry creation, fine meshing (requiring a lot of computer memory, especially at higher frequencies), and solving linear systems with matrix inversion routines, a simple 1-D model has been developed [13, 14]. This 1-D transverse plane wave approach can be used to obtain transfer matrix which is needed to cascade it with the preceding and succeeding elements constituting a complex muffler. The direction of the transverse plane wave is taken along the major-axis of the ellipse with the cavities above the inlet and beneath the outlet modeled as variable area resonators. The impedance of such a resonator is found using a semi-numeric technique called the matrizant approach [13]. Recently, this semi-analytical method has been replaced with an analytical method, where Frobenius solution of the differential equation governing the transverse plane wave propagation is obtained [14].

It may be noted from Fig. 5.47 that the flow has more than one parallel paths. Therefore, one needs to do a flow resistance network analysis and then evaluate the resistances of the perforated portions as well as the sudden expansions and contractions for use in the acoustic analysis [15].

In a multiply-connected muffler, waves in different perforated pipes have two sets of interactions with substantial frequency-dependent phase difference inbetween. This phenomenon is similar to that of Quincke tube (see Fig. 5.34). This results in acoustic interference which helps to raise the *TL* curve over a wider frequency range.

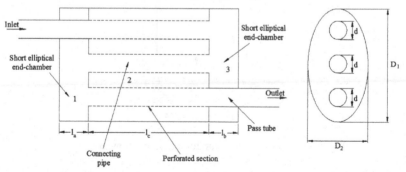

Fig. 5.47 Double-reversal end-chamber muffler.

Figure 5.47 shows only one of the many types of multiply connected mufflers. Some others are discussed by Panigrahi and Munjal [16].

## 5.8  Absorptive Ducts and Mufflers

An acoustically lined duct converts or dissipates acoustic energy into heat as a progressive wave moves along the duct, forward or backward. The lining can be a bulk reacting lining or locally reacting lining, depending on whether or not wave propagates through the body of the absorptive material. Common absorptive (or dissipative) materials are glass wool, mineral wool, ceramic wool, polyurethane (P.U.) foam, polyamide foam, etc. All these materials are highly porous with volume porosity of more than 95%. An acoustical material must have open (interconnected) pores unlike a thermal foam which has closed pores.

The most important physical parameter of absorptive material is flow resistivity. Referring to the schematic of a test setup shown in Fig. 5.48, flow resistivity, $E$, is given by

$$E = \frac{\Delta p}{Ud} \quad \left( \text{Pa.s} / \text{m}^2 \right) \tag{5.18}$$

where $\Delta p = \rho_w g \Delta h$ is the pressure drop across the sample of thickness $d$, and $U$ is the mean flow velocity in the test duct averaged over the cross-section.

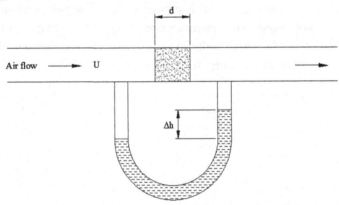

Fig. 5.48  Schematic of a setup for measuring flow resistivity.

Typically, flow resistivity E is of the order of 5000 – 40000 Pa.s/m$^2$. Generally, for typical thickness, $d$ (50 mm or 100 mm), the higher the value of $E$, the more will be value of TL of a lined duct. E increases with the mass density and fibre diameter (the smaller the fibre diameter, the more will be the value of $E$). An appropriate parameter is the non-dimensional acoustic flow resistance, $R$ defined as

$$R = \frac{Ed}{\rho_0 c_0} \tag{5.19}$$

Typically, $R$ is of the order of 1 to 8.

The most important design parameter is $h$, defined as

$$h \equiv \frac{\text{cross-section of the flow passage}}{\text{wetted (or lined) perimeter}} \tag{5.20}$$

Referring to Fig. 5.49, it may be noted that for a rectangular duct lined on two sides with an absorptive layer of thickness d, $h = b.2h / 2b$. Therefore, the flow passage height is denoted by '$2h$'. In other words, $h$ is half the flow passage height.

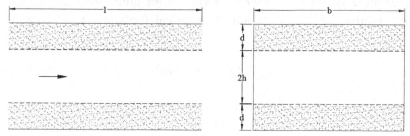

Fig. 5.49 Schematic of a rectangular duct lined on two sides.

Applying Eq. (5.20) to a circular duct lined all around with flow passage diameter $D$, yields

$$h = \frac{\pi/4 \, D^2}{\pi D} = \frac{D}{4} \tag{5.21}$$

Similar relationship would hold for a square duct lined on all four sides; $h$ would be one-fourth of the clear passage height. The most common practice in the heating, ventilation and air-conditioning (HVAC) systems is to make use of a rectangular duct lined on two sides, as shown in Fig. 5.49.

Transmission loss of a lined duct of length $l$ is given by

$$TL = TL_h . \, l \, / \, h \tag{5.22a}$$

where $TL_h$ is the specific $TL$ of a lined duct of length equal to $h$. This relationship is very helpful inasmuch as it indicates that $TL$ of a duct is directly proportional to its length $l$ and inversely proportional to h, defined by Eq. (5.20).

Equations (5.20) and (5.22a) collectively indicate that TL of a square duct lined on all four sides would be nearly double that of the duct lined on two opposite sides only.

Sometimes a designer or consultant needs to do some quick hand calculations of the effectiveness (in terms of transmission loss) of a lined duct. The datum available is $\bar{\alpha}$, the absorption coefficient of the material of the lining, defined as the fraction of the normally incident plane-wave energy absorbed by the given thickness of the lining, backed by a rigid wall. The value of the absorption coefficient $\bar{\alpha}$, supplied by the manufacturer, is an average value over a certain frequency range. There are a number of empirical formulae for quick hand calculations. One popular example is Piening's empirical formula [1], according to which

$$TL \approx 1.5 \frac{P}{S} \bar{\alpha} l \quad (dB) \tag{5.22b}$$

where $\bar{\alpha}$ is the absorption coefficient of the material,

$P$ is the lined perimeter, and

$S$ is the free-flow area of the cross section.

Thus, for a circular duct of radius $r_0$, or a square duct with each side $2r_0$ long, lined all over the periphery,

$$TL \approx 3\bar{\alpha}(l/r_0) \approx 3\frac{l}{d_0} \tag{5.22c}$$

Formula (5.22c) is indeed very useful for a quick estimate of the effectiveness of an acoustically lined duct. For example, it indicates that if a material with $\bar{\alpha} = 0.5$ were used to line a circular or square duct, it would yield a 3-dB attenuation across a length equal to one diameter or side length.

In order to reduce h (for increased $TL$ from a lined duct of given length $l$), we make use of a parallel baffle muffler shown in Figs. 5.19 and 5.30. These figures show the cross-section or end view of a 3-pass or 3-louvre parallel baffle muffler. It may be noted that the intermediate layer or baffle of thickness 2d would be servicing both the passages: the left half of it associated with the left passage and the right half with the right passage. Thus, a 3-louvre parallel baffle muffler is equivalent to three identical parallel rectangular ducts of the type shown in Fig. 5.49. Therefore, the *TL*-curves of Fig. 5.50 apply to a lined rectangular duct as well as a parallel baffle muffler.

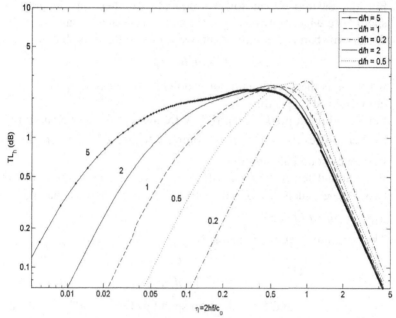

Fig. 5.50 Specific TL of a rectangular duct, or parallel baffle muffler, as a function of the non-dimensional frequency parameter for R =5 , illustrating the effect of d/h [4].

Specific TL parameter in Eq. (5.22a) may be read from Fig. 5.50, drawn on the lines of Ref. [4], where it is plotted against the non-dimensional frequency parameter, $\eta$, defined as

$$\eta = 2h f / c_0 = 2h / \lambda \qquad (5.23)$$

Parameter R in Fig. 5.50 is given by

$$R = \frac{Ed}{\rho_0 c_0} \qquad (5.24)$$

where *E* is flow resistivity of the absorptive filling.

Although the specific TL curves shown in Fig. 5.50 are applicable for R = 5, yet they can be used for other values of R (say, R = 1, 2, 10) without much error. Therefore, the curves of Fig. 5.50 can be used practically for most commercial absorptive materials made out of fibres, like glass wool, mineral wool, ceramic wool, foam, etc.

The following important observations may be made from Fig. 5.50:

1. For the desired *TL* at lower frequency (higher wavelength), we should use wider flow passages (larger $h$), and vice versa.

2. For narrow band *TL* requirement, we should select $h$ such that

$$2h/\lambda \approx 0.5 \; or \; h = \lambda/4$$

where $\lambda$ is the wavelength corresponding the peak frequency of the unmuffled SPL spectrum.

3. For absorbing a predominantly low frequency sound, we should use thicker baffles $(d > h)$. It is common to use $d = 2h$ for the low frequency sound attenuation.

4. If the unmuffled A-weighted SPL spans over several octaves, or over two decades, such that $\eta$ ranges from 0.02 to 2, then we may have to use $d = 5h \; (d \, / \, h = 5)$.

5. The open area fraction is given by

$$OAF = \frac{2h}{2h + 2d} = \frac{1}{1 + d/h} \tag{5.25}$$

Thus d/h = 1, 2 and 5 would correspond to OAF = 50%, 33.3% and 16.6 %, respectively.

6. Width of a square cross-section parallel baffle muffler with $n_p$ baffles (or, for that matter, n passages or louvres) is given by $W = 2(h + d)n_p$ in which $2hn_p$ represents the total flow passage width. Choice of a higher value of $d(d > h)$, within a fixed value of $W$, would decrease $h$ and/or $n_p$, and therefore carries the penalty of increased pressure drop or a larger parallel baffle muffler. Increased back pressure on the air handling fan (or blower) would
    a) reduce the air flow,
    b) call for a more powerful motor,
    c) increase its power consumption,
    d) increase the casing noise of the fan, and
    e) increase the aerodynamic noise generated in the louvers.

A larger parallel baffle muffler would
   a) need more space, and
   b) be heavier and costlier.
7. TL of an absorptive duct or parallel baffle muffler being proportional to its length presents an easy and sure way of increasing TL of an absorptive duct.

Observations 1 to 7 above indicate that optimum design of a parallel baffle muffler involves a compromise between acoustic attenuation, back pressure, size and cost [1, 4, 17].

Figure 5.50 applies to stationary medium. The convective effect of mean flow is to decrease *TL* marginally for the forward progressive wave, and increase it marginally for the rearward progressive wave (moving against the flow). This may, however, be neglected for a first design.

**Example 5.4** A tube-axial fan, operating in a tube of diameter *0.6 m* is producing sound power level in different octave frequency bands as follows.

| Frequency (Hz) | 63 | 125 | 250 | 500 | 1000 | 2000 | 4000 | 8000 |
|---|---|---|---|---|---|---|---|---|
| $L_w$ (dB) | 100 | 106 | 107 | 106 | 104 | 103 | 97 | 95 |

Calculate the space-averaged, A-weighted sound pressure level at 3m from the radiation end. If 4 m length of the 20 m long tube is lined on the inside with 100 mm thick blanket of highly porous polyurethane foam, calculate the overall transmission loss of the lining, and the attenuated $L_{PA}$ at the same measurement point outside the tube.

**Solution**

Making use of Table 1.2, the A-weighted sound power levels are tabulated below [see rows (b) and (c)].

For free field radiation, use of Eq. (4.6) gives the space-averaged, A-weighted SPL at 3 m:

$$L_{pA}(3\text{m}) = L_{WA} - 10 \log\left(4\pi \times 3^2\right)$$
$$= L_{WA} - 20.5$$

This is reflected in row (d) of the table below.

With 100 mm $(= 0.1\text{ m})$ lining inside the 0.6 m diameter tube, clear diameter is 0.4 m. Thus, the lining thickness $d = 0.1$ m, and making use of Eq. (5.20), $h = 0.4/4 = 0.1$ $m$. Therefore, $d/h = 1.0$, $l/h = 4/0.1 = 40$, and the non-dimensional frequency parameter $\eta$ in Fig. 5.50 is given by

$$\eta = \frac{2hf}{c_o} = \frac{2 \times 0.1}{346}\, f = 0.000578\, f$$

| Octave band center frequency (Hz) | 63 | 125 | 250 | 500 | 1000 | 2000 | 4000 | 8000 | Total |
|---|---|---|---|---|---|---|---|---|---|
| (a) $L_W$ (dB) | 100 | 106 | 107 | 106 | 104 | 103 | 97 | 95 | 112.9 |
| (b) A-weighting correction (dB) | − 26.2 | − 16.1 | − 8.6 | − 3.2 | 0.0 | 1.2 | 1.0 | − 1.1 | − |
| (c) $L_{WA} = (a) + (b)$ (dB) | 73.8 | 89.9 | 98.4 | 102.8 | 104.0 | 104.2 | 98.0 | 93.9 | 109.4 |
| (d) $L_{pA}(3\text{m}) = (c) - 20.5$ (dB) | 53.3 | 69.4 | 77.9 | 82.3 | 83.5 | 83.7 | 77.5 | 73.4 | 88.9 |
| (e) $\eta$ in Fig. 5.50 $= 0.000578\,f$ | 0.04 | 0.07 | 0.14 | 0.29 | 0.58 | 1.16 | 2.31 | 4.62 | − |
| (f) $TL_h$ from Fig. 5.50 for $d/h = 1$ (dB) | 0.3 | 1.0 | 1.3 | 2.2 | 2.5 | 1.5 | 0.3 | 0.07 | − |
| (g) $TL = TL_h \times l / h = (f) \times 40$ (dB) | 12 | 40 | 52 | 88 | 100 | 60 | 12 | 2.8 | − |
| (h) Attenuated $L_{pA}(3\text{m}) = (d) - (g)$ (dB) | 41.3 | 29.4 | 25.9 | − | − | 23.7 | 65.5 | 70.6 | 71.8 |

Thus, total attenuated $L_{pA}(3\text{m}) = 71.8$ dB [from row (h)], and overall transmission loss, $TL = 88.9 - 71.8 = 17.1$ dB [from rows (d) and (h)]

It is worth noting that for the data of Example 5.4 above (that is, the lined tube length, $l$ = 4 m, and clear diameter $d_0$ = 0.4 m), the grossly approximate relationship of Eq. (5.22c) suggests an overall TL of $3 \times 4/0.4 = 30$ dB, which is much more than the value of 17.1 $dB$, the end result of Example 5.4. However, it is very important that one selects $h$ and $d$ keeping in mind the unmuffled A-weighted SPL spectrum and Fig. 5.50. Often, additional considerations of the stagnation pressure drop, space availability and cost-effectiveness necessitate several iterations for the choice of $l$, $h$, $d$ and the number of passes (or louvres, or parallel baffles).

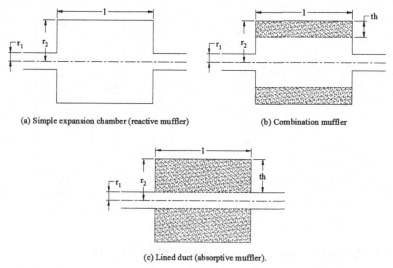

(a) Simple expansion chamber (reactive muffler)

(b) Combination muffler

(c) Lined duct (absorptive muffler).

Fig. 5.51 Schematics of muffler configurations illustrating the concept of a combination muffler $(r_2 = 3r_1)$.

## 5.9 Combination Mufflers

It may be observed from Fig. 5.50 that absorptive ducts and mufflers have the advantage of wide band attenuation over reactive mufflers, the *TL* curve of which may have sharp domes and troughs. On the other hand, the acoustic performance of absorptive ducts and mufflers is poor at low frequencies. Therefore, in order to raise the entire *TL* curve, particularly the troughs, it is desirable to make use of a muffler combining the two types of mufflers [18]. A simple configuration of a combination muffler is shown in Fig. 5.51(b) and the corresponding *TL* spectrum is shown in Fig. 5.52, where it is compared with *TL* of the limiting cases of a simple expansion chamber (reactive muffler) shown in Fig. 5.51(a) and a lined duct (absorptive muffler) shown in Fig. 5.51(c). Thickness of the absorptive lining has been used as a parameter in Fig. 5.52.

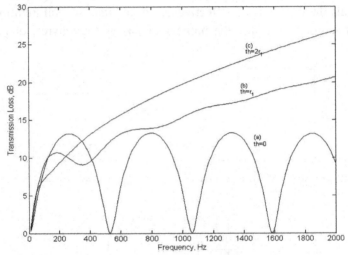

Fig. 5.52 Performance of the muffler configurations of Fig. 5.51.

## 5.10   Acoustic Source Characteristics of I.C. Engines

As per Thevenin theorem, analogous to electrical filter, the acoustic filter or muffler requires prior knowledge of the load-independent source characteristics $p_s$ and $Z_s$, corresponding to the open circuit voltage and internal impedance of an electrical source.

Prasad and Crocker [19], based on their direct measurements of source impedance of a multi-cylinder inline CI engine, proposed the anechoic source approximation: $Z_s = Y_0$. Callow and Peat [20] came out with a relatively more realistic expression:

$$Z_s(\text{exhaust}) = Y_0(0.707 - j0.707) \qquad (5.26a)$$

where $Y_0$ is the characteristic impedance of the exhaust pipe, $c_0/S$. Here, $S$ is the area of cross-section of the exhaust pipe, and $\rho_0$ and $c_0$ are density and the sound speed of the exhaust gases, respectively.

Hota and Munjal [21] extended the work of Fairbrother et al. [22,23] to formulate the source characteristics of a compression-ignition (CI) engine as functions of the engine's physical and thermodynamic parameters and incorporated them as empirical formulas into the scheme to predict the un-muffled noise using a multi-load method. Again, inspired by the work of Knutsson and Boden [24], the investigation of Ref. [24] was extended to the intake source characterization of C.I. engines by Hota and Munjal [25].

It has been found that internal impedance of the engine intake system is given by the empirical expression [25]

$$Z_s\left(\text{intake}\right) = Y_o\left(0.1 - j0.1\right) \qquad (5.26b)$$

Finally, Ref. [26] offers empirical expressions for the source strength level (SSL) in decibels of SI engines, for the intake as well as the exhaust system.

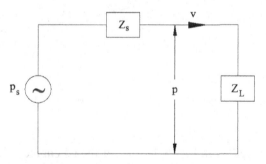

Fig. 5.53 Electrical analogous circuit for an un-muffled system [25].

A pre-requisite for this investigation is to have realistic values of the pressure-time history. These have been computed here making use of the commercial software AVL-BOOST [27] for different acoustic loads. This finite-volume CFD model is used in conjunction with the two-load method to evaluate the source characteristics at a point in the exhaust pipe just downstream of the exhaust manifold.

As per the electrical analogous circuit of the un-muffled system depicted in Fig. 5.53, for two different acoustic loads (impedances) $Z_{L1}$ and $Z_{L2}$, one can write:

$$p_s Z_{L1} - p_1 Z_s = p_1 Z_{L1} \quad \text{and} \quad p_s Z_{L2} - p_2 Z_s = p_2 Z_{L2} \qquad (5.27, 5.28)$$

These two equations may be solved simultaneously to obtain:

$$p_s = p_1 p_2 \frac{Z_{L1} - Z_{L2}}{p_2 Z_{L1} - p_1 Z_{L2}} \quad \text{and} \quad Z_s = Z_{L1} Z_{L2} \frac{p_1 - p_2}{p_2 Z_{L1} - p_1 Z_{L2}} \qquad (5.29, 5.30)$$

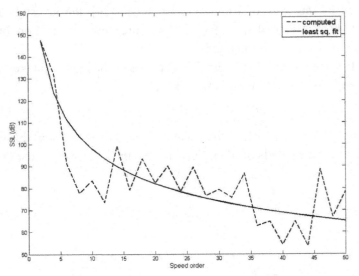

Fig. 5.54  Values of the intake SSL as a function of speed order for a turbocharged engine [25].

It may be observed from Fig. 5.54 that for the CI engines if a least square fit is done on the SSL spectrum at different frequencies or speed orders, the curve goes down more or less exponentially [21,25]. Hence the generalized formula for the SSL can be defined as:

$$SSL = A \times \left( \frac{\text{speed order}}{N_{cyl} / 2} \right)^B dB, \qquad (5.31)$$

where $N_{cyl}$ is the number of cylinders in the four-stroke cycle engine. $N_{cyl} / 2$ represents the speed order of the firing frequency of a four-stroke cycle engine, and constant A represents SSL at the firing frequency.

Speed order, $n$, of frequency $f_n$ is defined as

$$n = \frac{f_n}{RPM / 60} \qquad (5.32)$$

The firing frequency of a multi-cylinder engine is given by

$$\text{firing frequency} = \frac{RPM}{60} \times \frac{2}{N_{st}} \times N_{cyl} \qquad (5.33)$$

As there is one firing in two revolutions of a four-stroke ($N_{st} = 4$) cycle engine, the speed order of the firing frequency of a four-stroke cycle engine becomes $N_{cyl}/2$.

This kind of least square fit has been done to discount sharp peaks and troughs because computations have been made by assuming that speed of the engine remains absolutely constant. But in reality there may be around one to five percent variation in speed because the pressure-crank angle diagrams of successive cycles would never be identical.

Acoustic parametric study has been conducted for the following parameters, varying one at a time, keeping other parameters constant at their default (underlined) values:

**Turbocharged diesel engines**
Air fuel ratio, AFR = 18.0, **23.7**, 29.2, 38.0
Engine speed in RPM
= 1000, 1300, 1600, 2100, 2400, 3000, 3500, **4000**, 4500
Engine capacity (displacement), V (in liters)
= 1.0, 1.5, 2.0, **2.5**, 3.0, 4.0
Number of cylinders, $N_{cyl}$ =1, 2, 3, **4**, 6

So the default turbocharged engine is: 4 cylinders, 2.5 liters, running at 4000 rpm, with the air-fuel ratio 23.7.

**Naturally aspirated diesel engines**
Air fuel ratio, AFR = 14.5, **17.0**, 29.0, 39.6
Values of RPM, V and $N_{cyl}$ are the same as for the turbocharged engine above.

**A and B of SSL for Exhaust System of CI Engine [21]**
Turbocharged diesel engines:

$$A = 173.4 \times (1 - 0.0019\ AFR)(1 + 0.12\ NS - 0.016\ NS^2)$$
$$(1 - 0.0023\ V)(1 - 0.021\ N_{cyl})$$

$$B = -0.093 \times (1 + 0.016\ AFR)(1 + 0.31\ NS - 0.076\ NS^2)$$
$$(1 - 0.03\ V)(1 + 0.026\ N_{cyl})$$

(5.34a, b)

Naturally aspirated diesel engines:

$$A = 167 \times (1 - 0.0015\ AFR)(1 + 0.125\ NS - 0.021\ NS^2)$$
$$(1 + 0.0018\ V)(1 - 0.0233\ N_{cyl})$$

$$B = -0.13 \times (1 - 0.0123\ AFR)(1 + 0.19\ NS - 0.053\ NS^2)$$
$$(1 - 0.007\ V)(1 - 0.026\ N_{cyl})$$

(5.34c, d)

where $NS = engine\ speed\ in\ RPM/1000$.

**A and B of SSL for Intake System of CI Engine [25]**
Turbocharged diesel engines:

$$A = 214 \times (1 + 0.0018\ AFR)(1 - 0.08\ NS + 0.01\ NS^2)$$
$$(1 - 0.0021\ V)(1 - 0.05\ N_{cyl})$$

$$B = -0.318 \times (1 - 0.0033\ AFR)(1 - 0.039\ NS)$$
$$(1 - 0.173\ V)(1 + 0.022\ N_{cyl})$$

(5.35a, b)

Naturally aspirated diesel engines:

$$A = 189.6 \times (1 + 0.00075\ AFR)(1 - 0.1\ NS + 0.018\ NS^2)$$
$$(1 - 0.001\ V)(1 - 0.028\ N_{cyl})$$

$$B = -0.15 \times (1 + 0.0012\ AFR)(1 + 0.005\ NS)$$
$$(1 - 0.0064\ V)(1 + 0.109\ N_{cyl})$$

(5.35c, d)

where $NS = engine\ speed\ in\ RPM/1000$.

**Naturally aspirated SI engines [26]:**
An acoustic parametric study has been conducted for the following basic parameters, varying one at a time, keeping other parameters

constant at their default (underlined) values for both intake and exhaust.

Air fuel ratio, AFR = 12.0, **15.0**, 20.0, 24.0, 28.0

Engine speed in RPM = 1000, 1500, 2000, 2500, 3000, 3500, 4000, 4500, 5000, **5500**, 6000, 6500, 7000

Engine swept volume or displacement volume, V (in liters)
= 0.5, 1.0, 1.5, **2.0**, 2.5, 3.0,

Number of cylinders, $N_{cyl}$ = 1, 2, 3, **4**, 6

Here, default parameters of the naturally aspirated SI engine are: 4 cylinders, 2.0 liters, 5500 rpm, and the air-fuel ratio of 15. The resultant values of A and B of SSL are as follows:

Exhaust System:

$$A = 198.7 \times (1 - 0.0015\ AFR)(1 + 0.0445\ NS - 0.00765\ NS^2)$$
$$(1 - 0.0021\ V)(1 - 0.0374\ N_{cyl})$$

$$B = -0.053 \times (1 + 0.008\ AFR)(1 + 0.246\ NS - 0.0314\ NS^2)$$
$$(1 - 0.028\ V)(1 + 0.287\ N_{cyl}) \quad \text{(5.36a, b)}$$

Intake System:

$$A = 176.36 \times (1 - 0.00022\ AFR)(1 + 0.0517\ NS - 0.00866\ NS^2)$$
$$(1 - 0.00184\ V)(1 - 0.03336\ N_{cyl})$$

$$B = -0.106 \times (1 + 0.0021\ AFR)(1 + 0.256\ NS - 0.032\ NS^2)$$
$$(1 + 0.0075\ V)(1 + 0.122\ N_{cyl}) \quad \text{(5.37a, b)}$$

where $NS$ = *engine speed in RPM* /1000 .

The resultant source characteristics are used with the transfer matrix based muffler program [28] to predict the exhaust sound pressure level of a naturally aspirated four-stroke petrol or gasoline engine. Thus, the designer will be able to compute the exhaust sound pressure level with reasonable accuracy and thereby synthesize the required muffler configuration of a spark ignition (or gasoline) engine as well as the compression ignition (or diesel) engine.

## 5.11   Designing for Adequate Insertion Loss

As indicated before in Section 5.1, unlike transmission loss *TL*, insertion loss *IL* depends on the source impedance and radiation impedance as well as the overall transfer matrix of the muffler proper. While *TL* is appropriate for study of the inherent role of each element or a set of elements constituting the muffler proper, it is *IL* that is of interest to the user as a true measure of noise reduction due to a muffler. *IL* is a function of source impedance, radiation impedance, exhaust pipe, tail pipe as well as the muffler proper – see Fig. 5.2

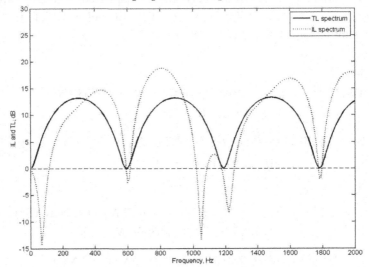

Fig. 5.55  Comparison of the IL and TL spectra for the simple expansion chamber
muffler configuration of Fig. 5.38.
$(c_0 = 592$ m/s, r = 2 cm, R = 6 cm, $l_e = 0.8$ m, l = 0.5 m, $l_t = 0.3$ m, M = 0).

The *IL* spectrum for a typical muffler is quite similar to the corresponding *TL* spectrum as shown in Fig. 5.55 for a simple expansion chamber muffler of Fig. 5.38. Yet there are extra dips or troughs because of the exhaust pipe and tail pipe resonances. The most important and disturbing feature of the *IL* spectrum is a prominent dip at low frequencies (typically, lower than 100 Hz). Here, *IL* is prominently negative indicating that at this frequency and its neighborhood, the exhaust *SPL* would be higher with muffler than that without it. In other words, at this frequency, the muffler would augment noise instead of muffling it! This low frequency dip (or dips) occurs for any source impedance (see, for example, Fig. 8.7 in Ref.1).

This effect and its remediation may be best illustrated by assuming the source to be a constant velocity source $(Z_s \to \infty)$. At such low frequencies, we may make use of the lumped element approximation ($k_0 l$, $k_0 l_e$ and $k_0 l_t$ being much less than unity) so that $\sin(k_0 l) \to k_0 l$, $\cos(k_0 l) \to 1$, etc., and radiation impedance $Z_0$ is much less than the tail pipe characteristic impedance $Y_t (= c_0/S_t)$. Then, the exhaust pipe and tail pipe behave like lumped inertance and the expansion chamber acts as lumped compliance (cavity), as shown in Fig. 5.56.

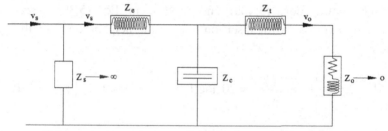

Fig. 5.56 Low-frequency (lumped-element) representation of the simple expansion chamber muffler of Fig. 5.38 for a constant velocity source and zero radiation impedance.

Making use of the linear network theory, it can easily be checked that

$$\frac{v_0}{v_s} = \frac{Z_c}{Z_c + Z_t} \quad (5.38)$$

where subscripts $e$, $c$ and $t$ denote exhaust pipe, chamber and tailpipe, respectively

Substituting Eq. (5.38) in Eq. (5.2) yields

$$IL = 20 \ \log \left| \frac{Z_c + Z_t}{Z_c} \right| \quad (5.39)$$

Equation (5.39) shows that $IL \to -\infty$ when $Z_c + Z_t \to 0$. This would happen when [1]

$$\frac{c_0^2}{j\omega V_c} + j\omega \frac{l_t}{S_t} = 0 \quad (5.40a)$$

where $V_c$ is volume of the chamber (cavity). Eq. (5.40a) yields the following frequency for the *IL* dip in Fig. (5.55)

$$f_{dip} = \frac{\omega}{2\pi} = \frac{c_0}{2\pi}\left(\frac{S_t}{V_c l_t}\right)^{1/2} \qquad (5.40b)$$

Substituting $c_0 = 592 \ m/s$ (corresponding to exhaust gas temperature of $600^0 C$, $S_t = \pi(0.02)^2 = 0.00126 m^2$, $V_c = \pi \times (0.06)^2 \times 0.5 = 5.655 \times 10^{-3} m^3$ (5.655 liters), Eq. (5.40b) indicates that the IL dip would occur at 81.1 Hz. It may be observed that despite the simplification/assumptions of lumped-element analysis and zero radiation impedance, this value is not far from that in Fig. 5.55.

Substituting Eqs. (5.40a) and (5.40b) in Eq. (5.39) yields the following expression for insertion loss in the neighborhood of the first dip:

$$IL \simeq 20 \ \log\left|1 - \omega^2 \frac{l_t V_c}{S_t c_0^2}\right| = 20 \ \log\left|1 - \left(\frac{f}{f_{dip}}\right)^2\right| \simeq 40 \ \log\left(f/f_{dip}\right) \quad (5.41)$$

Thus, at $f = 2f_{dip}$, one can expect an IL of about 9.5 dB, as may indeed be verified from Fig. 5.55.

Significantly, the approximate expressions of Eqs. (5.40) and (5.41) throw up the following design guidelines for the engine exhaust muffler, where $f_{dip}$ must be sufficiently less than the firing frequency of the engine:

1. The tail pipe should be long and of narrow cross-section, within the constraints of space and the overriding requirements of low back pressure and aerodynamic noise generation.
2. The muffler proper should have sufficiently large volume. This requirement is reinforced by the need to have large expansion ratio (sharp impedance mismatch). Table 5.1, reproduced here from Refs. [29, 30], gives an idea of how important the muffler volume could be in different octave frequency bands, not just at the low frequency dip. As a thumb rule, small, medium and large mufflers are roughly characterized by 5, 10 and 15 times the piston displacement capacity of the engine, respectively.

3. We should ensure that exhaust (and intake) system runners are of the same acoustical length so that the firing frequency of the engine is equal to $n_{cyl}$ times that of a single cylinder, where $n_{cyl}$ denotes the number of cylinders — see Eq. (5.33). This would ensure that $f_{dip}$ is much less than (say one-half of) the firing frequency of the engine. If the runner lengths are not equal, as indeed was the case in the older engine designs, then the unmuffled exhaust noise spectrum would show additional low frequency peaks at the single-cylinder firing frequency (RPM/60 for the 2-stroke cycle engines and RPM/120 for the 4-stroke cycle engines). For such an engine, the muffled exhaust noise would sound considerably louder and harsher (of poor sound quality).

Table 5.1. Approximate insertion loss (dB) of typical reactive mufflers used with reciprocating engines[a] [29].

| Octave band Centre Frequency (Hz) | Low pressure-drop muffler | | | High pressure-drop muffler | | |
|---|---|---|---|---|---|---|
| | Small | Medium | Large | Small | Medium | Large |
| 63 | 10 | 15 | 20 | 16 | 20 | 25 |
| 125 | 15 | 20 | 25 | 21 | 25 | 29 |
| 250 | 13 | 18 | 23 | 21 | 24 | 29 |
| 500 | 11 | 16 | 21 | 19 | 22 | 27 |
| 1000 | 10 | 15 | 20 | 17 | 20 | 25 |
| 2000 | 9 | 14 | 19 | 15 | 19 | 24 |
| 4000 | 8 | 13 | 18 | 14 | 18 | 23 |
| 8000 | 8 | 13 | 18 | 14 | 17 | 23 |

[a] Refer to manufacturers literature for more specific data.

**Example 5.5** For a 500 cc, two-cylinder four-stroke cycle engine, directly driving an agricultural pump at 3000 RPM, design a single simple expansion chamber muffler (find its volume and the tail pipe diameter and length) so as to ensure an insertion loss of at least 10 dB at the firing frequency of the engine.

**Solution**

Equations (5.40) and (5.41) indicate that we need to provide large enough expansion chamber volume $V_c$ commensurate with the engine capacity as per Table 5.1, and tail pipe with small enough cross-sectional area $S_t$ and large enough length $l_t$.

The smallest cross-section of the tail pipe is decided by the consideration of the mean flow Mach number not exceeding 0.2. This would need prior knowledge of the volumetric efficiency and air-fuel ratio of the engine, apart from its capacity, RPM, number of cylinders, cycle-strokes, etc.

For the present problem, let us assume as follows:

$V_c$ = 10 times the engine capacity = $10 \times 500/10^6$ = $0.005\ m^3$

tail pipe diameter = $2\ cm \Rightarrow S_t = \dfrac{\pi}{4} \times \left(\dfrac{2}{100}\right)^2$ = $3.1416 \times 10^{-4}\ m^2$

exhaust gas temperature = $600^o C$, so that sound speed $c_0 = 592\ m/s$.

For a 2-cylinder four-stroke cycle engine running at 3000 RPM, the firing frequency is given by:

$$\omega = 2\pi \frac{3000}{60 \times 2} \times 2 = 314.16\ rad/s$$

Substituting these data in Eq. (5.41) yields

$$10 = 20 \log\left|1 - (314.16)^2 \frac{l_t \times 0.005}{3.1416 \times 10^{-4} \times (592)^2}\right|$$

which gives the required tail pipe length: $l_t = 0.928\ m$, which is reasonable. Thus, an exhaust system with

$$V_c \geq 0.005\ m^3 = 5\ litres, \quad d_t \leq 2\ cm \quad \text{and} \quad l_t \geq 0.928\ m$$

would ensure an insertion loss of at least 10 dB at the firing frequency of the given engine.

It may be noted from Eq. (5.41) that insertion loss would improve at lower exhaust temperature and/or with larger expansion chamber or cavity volume.

## 5.12   Mufflers for High Pressure Vents and Safety Valves

The foregoing sections have dealt with the intake and exhaust systems of reciprocating engines, fans, blowers, HVAC systems, etc. where the static pressure inside the duct is nearly atmospheric, flow is incompressible, and sound is produced by reciprocating motion of the piston, temporal variation of flow passages and/or periodic cutting of air by rotating blades.  By contrast, in the high pressure vents and safety valves, the static pressure is much above (and often several times) the atmospheric pressure, flow is compressible (choked sonic at the valve throat), and sound is produced by the high velocity flow of air, gases or superheated steam. High pressure superheated steam is produced by boilers that take days to stabilize.  If, for whatever reasons or eventuality, the steam is not needed for some time, then the boiler cannot be stopped; it keeps producing high pressure superheated steam, which has to be vented out to the atmosphere. And that produces very high aerodynamic noise.

Another eventuality occurs when due to a malfunction in the line, pressure builds up in the receiver (or plenum) upstream. At a pre-set pressure, a safety valve (or pressure relief valve) opens in order to relieve the extra pressure. This too produces high aerodynamic noise.

To minimize the radiation of aerodynamic noise and absorb the noise so produced we use multi-stage diffuser and parallel baffle muffler, respectively.  Use of a couple (or more) of multi-hole orifice plates downstream of the valve reduce the high pressure upstream of the valve (in the receiver) in stages.  This helps in several ways as follows.

1. Multi-stage expansion (the valve is the first stage in the series) produces much less aerodynamic (flow) noise than a single stage (the valve only).
2. The aerodynamic noise produced by one stage is substantially absorbed by expansion in the following stage(s). Thus, the noise

produced at the last stage only needs to be addressed by the parallel baffle muffler.

3. The peak frequency of aerodynamic noise scales with Strouhal number, defined by

$$N_s = \frac{f_p d}{U} \approx 0.2 \implies f_p \approx \frac{0.2U}{d} \qquad (5.42)$$

where $N_S$ is Strouhal number, $f_p$ is the peak frequency of the wide-band aerodynamic noise spectrum, $U$ is velocity of the jet emerging from each hole and d is diameter of the hole. As $U$ is near sonic (~345 m/s) and $d = 5$ *mm* (or 10 *mm*), the peak frequency would be of the order of 13800 (or 6900) *Hz*. To absorb such high frequency noise, the required parallel baffle muffler would be of modest dimensions because a parallel baffle muffler is much more efficient at higher frequencies.

Schematic of a typical configurations of multi-stage diffuser-cum-parallel baffle muffler is shown in Fig. 5.57.

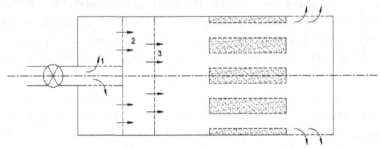

Fig. 5.57 Schematic of a 3-stage diffuser-cum-parallel baffle muffler.

An alternative to multi-stage diffuser consisting of multi-hole orifice plates is the use of a multi-expansion trim within the control valve. Some of these high-tech proprietary multiple-step trims are given in Ref. [31] and are reproduced in the next chapter (sec. 6.11). As the pressure decreases, specific volume increases and therefore the required total throat area would increase with every stage. So, a larger number of holes would be needed as one moves downstream. This calls for a Computational Fluid Dynamics (CFD) analysis of the flow through all the expansion stages including the valve in order to evaluate static pressure, specific volume and the number of holes required at each stage. For prediction of noise radiated by control valves, the reader is referred to Baumann and Hoffman [32], or to Bies and Hansen [30].

## 5.13 Design of Muffler Shell and End Plates

In nearly all the muffler configurations discussed above, it is presumed that the muffler shell and end plates are rigid or their compliance is zero. In actual practice, however, their compliance is non-zero. Excited by the acoustic pressure fluctuations, the shell as well as end-plates are set into vibration and radiate secondary sound outside. This is called the break-out noise or the shell noise, and is illustrated in Fig. 5.58. This adds up to the noise radiated axially through the tail pipe of the muffler and the net *TL* is then less than the axial TL as shown below.

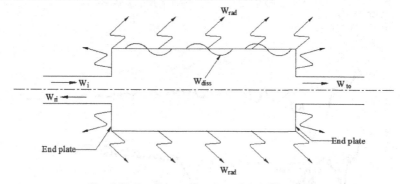

Fig. 5.58 Break-out noise from the muffler shell.

Referring to Fig. 5.58, and neglecting dissipation of power $(W_{diss} \to o)$,

$$\text{Axial transmission loss, } TL_a = 10 \, \log(W_i/W_{to}) \qquad (5.43)$$

$$\text{Transverse transmission loss, } TL_{tp} = 10 \, \log(W_i/W_{rad}) \qquad (5.44)$$

$$\text{Net transmission loss, } TL_{net} = 10 \, \log(W_i/(W_{to} + W_{rad})) \qquad (5.45)$$

where $W_i$ is the incident acoustic power in the exhaust pipe,

$W_{ri}$ is the reflected acoustic power in the exhaust pipe,

$W_{to}$ is the acoustic power transmitted into an anechoic termination along the tail pipe,

$W_{rad}$ is the acoustic power radiated as break-out noise from the shell and end plates, and

$W_{diss}$ is the power that is dissipated into heat through viscosity or structural damping.

Substituting Eqs. (5.43) and (5.44) into Eq. (5.45) yields

$$TL_{net} = -10 \, \log\left(10^{-0.1 \, TL_a} + 10^{-0.1 \, TL_{tp}}\right) \tag{5.46}$$

Equation (5.46) shows that $TL_{net}$ at any frequency would be less than the lower of $TL_a$ and $TL_{tp}$ by 0 to 3 $dB$.

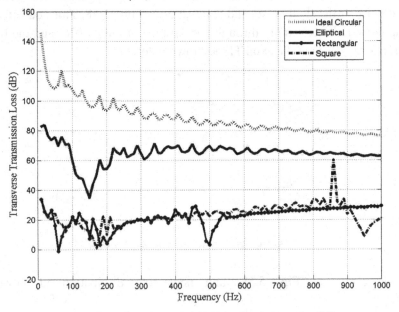

Fig. 5.59 Effect of non-circularity on break-out noise [34].

Unlike axial transmission loss, $TL_a$, which is zero as frequency f tends to zero, the transverse transmission loss, $TL_{tp}$ tends to infinity as f tends to zero, as is typical of the stiffness controlled $TL$. $TL_{tp}$ is maximum for an ideal circular cylinder and falls drastically with non-circularity [33], as is shown in Fig. 5.59. Deviation from circularity decreases the transverse transmission loss (TTL) and hence increases the breakout noise substantially [34]. That is why TTL of rectangular (or square) ducts is the lowest [35], which accounts for 'cross talk' in the HVAC systems. This indicates the importance of proper design of shell and end plates [36] as well as the inner acoustic elements of the muffler.

The break-out noise from the end plates may be reduced by stiffening them and giving them a dish shape; flat plates act as efficient radiators. Offset inlet and outlet automatically serve to stiffen the end plates and should be preferred in order to reduce the break-out noise.

## 5.14 Helmholtz Resonators

The resonator developed by Helmholtz is as old as the science of Acoustics. It consists of a cavity (lumped volume) attached to a duct through a small narrow tube (neck) as shown in Fig. 5.60(a). Its analogous circuit presentation is shown in Fig. 5.60(b), where it is represented as a branch or shunt impedance consisting of a lumped inertance due to the neck and compliance due to the cavity. Additionally, there is radiation impedance at both ends of the neck. Thus, the impedance of the resonator is given by [1]

$$Z = \frac{\omega^2}{\pi c_0} + j\omega \, \frac{l_{ne}}{S_n} + \frac{c_0^2}{j\omega \, V_c} \tag{5.47a}$$

where $S_n$ is the neck cross-sectional area, $\pi \, d_n^2/4$
$l_{ne}$ is the effective length of the neck given by

$$l_{ne} = l_n + t_w + 0.85 d_n \tag{5.47b}$$

Here, the end correction $0.85 d_n$ represents the sum of the inertive parts of the radiation impedance at the two ends of the neck.

At the resonance frequency of the resonator, the reactive (imaginary) part of $Z$ would tend to zero. Thus, the resonance frequency of Helmholtz resonator is given by

$$j\omega \, \frac{l_{ne}}{S_n} + \frac{c_0^2}{j\omega \, V_c} = 0$$

which, on rearrangement, yields

$$f_n = \frac{\omega}{2\pi} = \frac{c_0}{2\pi}\left(\frac{S_n}{l_{ne}V_c}\right)^{1/2} \qquad (5.48)$$

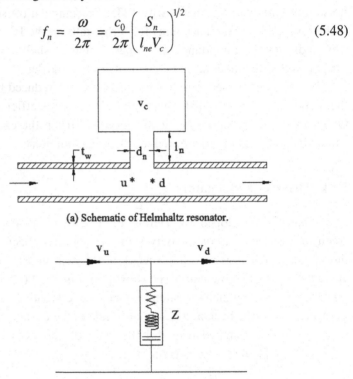

(a) Schematic of Helmhaltz resonator.

(b) Analogous circuit of Helmholtz resonator

Fig. 5.60 Lumped-element modeling of Helmholtz resonator in a duct.

At this frequency, most of the incoming acoustic power would be reflected back to the source because of the acoustical short-circuit. At this frequency, TL of the Helmholtz resonator would tend to infinity. Thus, the TL curve would show a sharp peak —— the resonance peak of the Helmholtz resonator.

The half-power bandwidth of the resonance peak in the *TL* curve would tend to zero if the resistive part of the resonator impedance in Eq. (5.47) were zero. However, the term $\omega^2/\pi c_0$ gives a finite width and amplitude to the resonance peak. The half-power bandwidth can be increased at the cost of its amplitude by lining the cavity acoustically. The acoustical as well as anti-drumming lining on the inside of the cavity helps to reduce the structural vibration of the cavity and the consequent break-out noise.

Helmholtz resonators are characterized by high TL at the resonance frequency. They are ineffective at all other frequencies except in the immediate neighborhood of the resonance frequency. Therefore, they find application on constant speed machines like generator sets (gensets). They are also used sometimes on the intake systems of automotive engines in order to suppress a boom in the passenger cabin at the idling speed of the engine.

Another application of Helmholtz resonator or its uniform-area transverse tube variant is on the intake and/or discharge duct of the constant-speed fans or blowers. As the unmuffled intake and exhaust-noise spectra are characterized by sharp peaks at the blade passing frequency (BPF) and its first few multiples, one can design and employ transverse tubes resonating at the BPF, 2 times BPF and 3 times BPF as shown in Fig. 5.61.

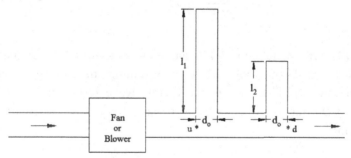

(a) Schematic of the configuration with two resonators.

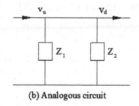

(b) Analogous circuit

Fig. 5.61 Transverse tube resonators for the exhaust noise control of a constant speed fan or blower.

The branch impedance of transverse uniform tube of length $l$ and area of cross-section $S$ is given by

$$Z = -jY_0 \cot k_0 l, \quad Y_0 = c_0/S \tag{5.49}$$

This would tend to zero when

$$k_0 l = (2m-1)\pi/2, \quad k_0 = \omega/c_0 = 2\pi f_m/c_0 \tag{5.50a}$$

or when the frequency equals one of the resonance frequencies:

$$f_m = (2m-1)\frac{c_0}{4l}, \quad m = 1,2,3,\ldots\ldots\ldots\ldots \tag{5.50b}$$

Thus, if $l = c_0/(4 \times BPF)$, then it would suppress the BPF peak and its odd multiples. Similarly, a transverse tube of half the length would suppress the SPL peaks occurring at even multiples of the BPF. Hence, referring to Fig. 5.61, and incorporating the end corrections, the design values of $l_1$ and $l_2$ work out to be as follows:

$$l_1 = \frac{c_0}{4 \times BPF} - 0.425\, d_0 \tag{5.51}$$

$$l_2 = \frac{c_0}{8 \times BPF} - 0.425\, d_0 \tag{5.52}$$

**Example 5.6** A tube-axial fan has eight blades and is running at a constant speed of 1500 RPM. Assuming the average ambient temperature of $25^0\ C$, design the quarter-wave resonators in order to suppress the fan noise at its blade passing frequency and its first two harmonics.

**Solution**

Blade passing frequency of the fan, $BPF = \dfrac{8 \times 1500}{60} = 200$ Hz

At $25^0\ C$, source speed $c_0 = 346$ m/s

Let the transverse quarter-wave resonator tube diameter, $d_o = 0.1$ m

Then, making use of Eqs. (5.51) and (5.52), the required lengths of the resonators (see Fig. 5.61) are given by

$$l_1 = \frac{346}{4 \times 200} - 0.425 \times 0.1 = 0.39 \text{ m} = 390 \text{ mm}$$

$$l_2 = \frac{346}{8 \times 200} - 0.425 \times 0.1 = 0.174 \text{ m} = 174 \text{ mm}$$

The resonator tube of 390 mm length would absorb BPF and its $3^{rd}$, $5^{th}$,…. multiples, and the tube of 174 mm length would absorb $2^{nd}$, $6^{th}$, ............ multiples of the BPF. Therefore, two resonator tubes of length 390 mm and 174 mm would suffice to suppress the BPF and its first two harmonics.

There is another interesting application of Helmholtz Resonators (HR). When a large number of identical HR's are applied over a wall, they can make the wall anechoic to plane wave incident on the wall when the plane wave frequency equals the resonance frequency of the HR's. At resonance, impedance Z of Eq. (5.47) reduces to pure resistance, $R = \omega^2/\pi c_0$. This resistance would absorb acoustic power equal to $p_{rms}^2/(\rho_0 R)$. Now, power in a plane wave normal to area A is $p_{rms}^2 A / \rho_0 c_0$. Equating the two gives an expression for area A over which the wall will act as anechoic. Thus,

$$\frac{p_{rms}^2 A}{\rho_0 c_0} = \frac{p_{rms}^2}{\rho_0 R} \qquad (5.53)$$

or

$$A = \frac{c_0}{R} = \frac{c_0}{\omega^2/\pi c_0} = \frac{\pi c_0^2}{\omega^2} \approx \frac{0.953}{(f/100)^2} \left(\text{m}^2\right) \qquad (5.54)$$

Thus, at room temperature of $25^0 C \left(c_0 = 346 \ m/s\right)$, an HR resonating at 100 Hz would make 0.953 $m^2$ area of the wall anechoic at that frequency.

## 5.15 Active Noise Control in a Duct

An active noise control system makes use of a secondary sound source and an adaptive digital controller to generate a pressure signal that is out of phase with that of the primary sound source. Schematic of an active noise control system in a duct is shown in Fig. 5.62. Signal conditioner shown there may consist of a preamplifier, low-pass filter (LPF) and analog-to-digital (A/D) converter. Secondary source of sound is generally a loudspeaker specially designed for the purpose. The error signal is used to adapt the filter of the microprocessor in order to generate the required transfer function that would help to minimize acoustic pressure at the error microphone.

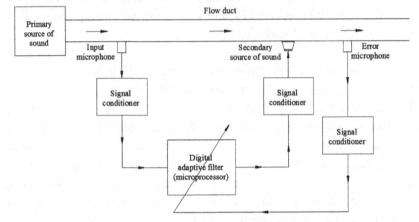

Fig. 5.62 Schematic of an active noise control system in a duct.

The acoustic signal produced by the secondary source travels upstream as well as downstream, and is picked up (sensed) by the input microphone as well as the error microphone. It has been proved by means of a 1-D standing wave analysis [37] as well as the progressive wave transfer function analysis [38] that in an active noise control system in its steady state, the primary source faces nearly zero acoustic resistance, and so does the secondary source. In other words, the two sources unload each other. The controller helps to create an acoustical short circuit at the secondary source junction. When the acoustic load (resistance) at each source is zero, the particle (or volume) velocity is out of phase with pressure, and therefore, does not produce any audible sound. In other words, the primary source and secondary source muffle each other. Thus, an active noise control (ANC) system produces a sharp impedance mismatch within a limited frequency band [37-41].

Adaptive controllers are basically of two types: feedforward controller and feedback controller. Feedforward controller works efficiently for periodic noise (from fans, blowers, HVAC systems that run at constant speed) or for random (turbulence generated) noise propagating in ducts. The feedback controller works best for active earmuffs (headsets), active vehicle suspension systems and active control of structural vibration. Analysis and design of controllers call for extensive knowledge of digital signal processing (DSP), which is out of the scope of the text book.

A typical engine exhaust system or industrial fan produces a signal that varies with time because of small (but unavoidable) variations in speed, and therefore adaptive controller is a necessity for an ANC system in a duct.

Large (powerful) loudspeaker(s), signal conditioners and adaptive DSP controller make an active noise control a costly proposition. However, such a system has the advantage of minimal stagnation pressure drop which reduces the running cost of the system dramatically. Thus, for large industrial fans, the active noise control system reduces the power bill substantially. It not only offsets the disadvantage of high initial hardware cost, but also offers a very cost effective alternative to passive noise control system.

ANC system in a duct works best for plane wave propagation. The presence of higher order modes at high frequencies would call for more than one input microphone, secondary source loudspeakers and error phones, apart from a more complex DSP controller. This would make the ANC system uneconomical at higher frequencies. For the same reason, ANC system is uneconomical for use in 3-D enclosures or rooms for global noise control. However, ANC system can be used on seats of luxury vehicles, aircraft passenger cabins, etc. for localized noise control, roughly within a radius of quarter wavelength.

ANC system is most effective at lower frequencies (lower than 1000 Hz, rather 500 Hz) where passive noise control system, particularly the absorptive type, is not very effective. This makes the active noise control system complementary to the passive noise control system for wide-band noise control requirements.

The ANC system, when tuned adaptively, locks on to the peaks in the unmuffled noise spectrum, and chops them off, as it were. This is what makes the ANC system particularly suitable for large fans and blowers, the noise spectrum of which is characterized by sharp peaks at the blade passing frequency and its first few harmonics.

## 5.16  Pressure Drop Considerations

Mean flow through the muffler passages encounters a drop in its stagnation pressure. The engine or fan source must push the gases or air against this back pressure. This results in pumping losses in the engine power output. When back pressure is high, it results in a drop in volumetric efficiency, which in turn would result in a further loss in engine shaft power and an increase in the specific fuel consumption (SFC) of the engine. In the case of a fan, the power load on the motor is proportional to the product of the flow rate $Q$ and total pressure drop, $\Delta p$. The latter is the sum of the pressure drops in the intake system and discharge/exhaust system. Thus, electrical power consumption of a fan motor is directly proportional to the total back pressure experienced by the fan. Therefore, prediction and minimization of the stagnation pressure drop across mufflers (or silencers) is an integral part of the discipline of analysis and design of mufflers.

Two primary mechanisms of the stagnation pressure drop are wall friction in the boundary layer, and eddy and vortex formation in the free shear layer at area discontinuities and perforated sections.

Pressure drop can be normalized in terms of dynamic head, $H$. Thus,

$$\Delta p = K.H, \quad H \equiv \frac{1}{2}\rho_0 U^2 \qquad (5.55)$$

Here, $K$ is called the dynamic pressure loss factor or coefficient. Empirical expressions of $K$ for different muffler elements with turbulent mean flow are given below.

Pressure loss factor for turbulent flow through a pipe or duct of length $l_p$ and (hydraulic) diameter $d_p$ (Fig. 5.3):

$$K \equiv \frac{\Delta p}{H} = F.\frac{l_p}{d_p} \qquad (5.56)$$

where $F$ is the Froude's friction factor given by

$$F = 0.0072 + 0.612 / R_e^{0.35}, \quad R_e < 4 \times 10^5 \quad (5.57)$$

$R_e = U d_p \rho_0 / \mu$ is the Reynold's number, and $\mu$ and $\rho_o$ are the coefficient of dynamic viscosity and mass density, respectively

Typically, $F \approx 0.016$ for the mild steel pipes used in exhaust mufflers. However, roughness of fiberglass-lined pipes is much more than that of the normal unlined rigid pipes, and therefore, $F \approx 0.032$ for the lined pipes as well as the perforated unlined pipes, as in a concentric tube resonator (Fig. 5.10).

Pressure loss factor for a sudden contraction (Fig. 5.4):

$$K \approx (1-n)/2, \quad n = (r_d / r_u)^2 \quad (5.58)$$

Pressure loss factor for a sudden expansion (Fig. 5.5):

$$K \approx (1-n)^2, \quad n = (r_u / r_d)^2 \quad (5.59)$$

Note that for both the sudden area discontinuities, $n$ is defined as ratio of cross-section of the narrower pipe to that of the wider pipe. Thus, $n$ is less than unity in both Eqs. (5.58) and (5.59).

Pressure loss factor for extended outlet (Fig. 5.6), extended inlet (Fig. 5.7), flow reversal with contraction or expansion:

$$K \approx 1.0 \quad (5.60)$$

Pressure loss factors for bell mouth, gradual contraction, gradual expansion, etc. are much lower than those for the corresponding sudden or step area changes. However, gradual area changes lead to poor *TL* and therefore are not always desirable.

Pressure loss coefficient of the plug muffler (Fig. 5.44) [12]:

$$K = 5.6e^{-0.23x} + 67.3e^{-3.05x}, \quad 0.25 < x < 1.4 \quad (5.61)$$

where $x$ is the open-area ratio of the cross-flow expansion (Fig. 5.11) and cross-flow contraction (Fig. 5.12), assumed to be equal. The open-area ratio is defined by

$$\text{open-area ratio,} \quad x = \frac{2\pi r_1 l_p \sigma}{\pi r_1^2} = \frac{2 l_p \sigma}{r_1} \quad (5.62)$$

where $\sigma$ is porosity of each of the two perforated sections.

Pressure loss coefficient of the muffler element with three interacting ducts (Fig. 5.13) [12]:

$$K = 4.2e^{-0.06x} + 16.7e^{-2.03x} \qquad (5.63)$$

By contrast, pressure loss coefficient of a concentric tube resonator (Fig. 5.10) is as small as [12]

$$K = 0.06\,x \qquad (5.64)$$

where $x$ is the open-area ratio defined by Eq. (5.62).

**Example 5.7**  For length $l = 0.5\ m$, $r = 2\ cm$, $R = 6\ cm$ estimate the stagnation pressure drop across the simple expansion chamber muffler of Fig. 5.38.

**Solution**

Area ratio, $n = (r/R)^2 = (2/6)^2 = 0.111$

As per Eq. (5.59), the pressure loss coefficient for sudden expansion, $K_e = (1-n)^2 = (1-0.111)^2 = 0.790$

As per Eq. (5.56), pressure loss coefficient for wall friction to turbulent flow through the chamber length,

$$K_f = Fl / D \simeq 0.016\frac{0.5}{2\times0.06} = 0.067$$

As per Eq. (5.58), pressure loss coefficient for sudden contraction,

$$K_c = (1-n)/2 = (1-0.111)/2 = 0.444$$

Let the dynamic head in the exhaust (and tail) pipe, $H = \dfrac{1}{2}\rho_0 U^2$

Then, flow velocity in the chamber $= U \times n = 0.111\ U$
And therefore, dynamic head in the chamber

$$= (0.111)^2 H = 0.0123\ H$$

Thus, the total stagnation pressure drop across the muffler proper (excluding the exhaust pipe and tail pipe) is given by

$$\Delta_p = (0.790 + 0.0123\times0.067 + 0.444)\ H$$

$$= (0.790 + 0.0008 + 0.444) \, H = 1.235 \, H$$

Hence, the total stagnation pressure drop is equal to 1.235 times the dynamic head in the exhaust/tail pipe.

It may be noted that the contribution of the chamber to the total pressure drop is very small, and therefore can be neglected for most design calculations.

For multiply-connected perforated element mufflers, one needs to make use of the lumped flow resistance network circuit [15] in order to predict the pressure loss coefficient of the muffler. Fortunately, suchlike mufflers are characterized by quite small pressure drop, and therefore represent a viable design option.

The foregoing expressions for the pressure loss coefficient are approximate. Therefore, ultimately, the overall back pressure of the muffler has to be measured on the test bed.

In the case of automotive engines, the after-treatment devices like catalytic converter, diesel particulate filter, etc., add substantially to the overall back pressure experienced by the engine. That leaves only a modest pressure drop for which the muffler needs to be designed. In fact, the maximum permissible pressure drop across the muffler is the most important consideration for synthesizing an appropriate muffler configuration.

## References

1. Munjal, M. L., Acoustics of Ducts and Mufflers, John Wiley, New York, (1987).
2. Munjal, M. L., Muffler Acoustics, Chap. K in F.P. Mechel (Ed.), Formulas of Acoustics, Springer Verlag, Berlin, (2006).
3. Munjal, M. L., Sreenath, A. V. and Narasimhan, M. V., A rational approach to the synthesis of one-dimensional acoustic filters, Journal of Sound and Vibration, 29(3), pp. 263-280, (1973).
4. Munjal, M. L., Galaitsis, A. G. and Ver, I. L., Passive Silencers, Chap. 9 in I. L. Ver and L. L. Beranek (Ed.) Noise and Vibration Control Engineering, John Wiley, New York, (2006).

5. Hans Boden, and Ragnar Glav, Exhaust and Intake Noise and Acoustical Design of Mufflers and Silencers, Chap. 85 in M.J. Crocker (Ed.) Handbook of Noise and Vibration Control, John Wiley, New York, (2007).

6. Kinsler, L. E., Frey, A. R., Coppens, A. B. and Sanders, J. V., Fundamentals of Acoustics, Fourth Edition, John Wiley, New York, (2000).

7. Chaitanya, P. and Munjal, M. L., Effect of wall thickness on the end-corrections of the extended inlet and outlet of a double-tuned expansion chamber, Applied Acoustics, 72(1), pp. 65-70, (2011).

8. Karal, F.C., The analogous impedance for discontinuities and constrictions of circular cross –section, Journal of the Acoustical Society of America, 25(2), pp. 327-334, (1953).

9. Sahasrabudhe, A. D., Munjal, M. L. and Ramu, P. A., Analysis of inertance due to the higher order mode effects in a sudden area discontinuity, Journal of Sound and Vibration, 185(3), pp. 515-529, (1995).

10. Sullivan, J. W. and Crocker, M. J., Analysis of concentric tube resonators having unpartitioned cavities, Journal of the Acoustical Society of America, 64, pp. 207-215, (1978).

11. Chaitanya, P. and Munjal, M. L., Tuning of the extended concentric tube resonators, International Journal of Acoustics and Vibration, 16(3), pp. 111-118, (2011).

12. Munjal, M. L., Krishnan, S. and Reddy, M. M., Flow-acoustic performance of the perforated elements with application to design, Noise Control Engineering Journal, 40(1), pp. 159-167, (1993).

13. Mimani, A. and Munjal, M. L., Transverse plane-wave analysis of short elliptical end-chamber and expansion chamber mufflers, International Journal of Acoustics and Vibration, 15(1), pp. 24-38, (2010).

14. Mimani, A. and Munjal, M. L., Transverse plane-wave analysis of short elliptical chamber mufflers – an analytical approach, Journal of Sound and Vibration, 330, pp. 1472-1489, (2011).

15. Elnady, T., Abom, M. and Allam, S., Modeling perforate in mufflers using two-ports, ASME Journal of Vibration and Acoustics, 132(6), pp. 1-11, (2010).

16. Panigrahi, S. N. and Munjal, M. L., Plane wave propagation in generalized multiply connected acoustic filters. Journal of the Acoustical Society of America, 118(5), pp. 2860-2868, (2005).

17. ASHRAE (American Society of Heating, Refrigeration and Air-Conditioning Engineers) Handbook: HVAC Applications, Chapter 46, (1999).

18. Panigrahi, S. N. and Munjal, M. L., Combination mufflers --- theory and parametric study, Noise Control Engineering Journal, 53(6), pp. 247-255, (2005).

19. Prasad, M. G. and Crocker, M. J., On the measurement of the internal source impedance of a multi-cylinder engine exhaust system, Journal of Sound and Vibration, 90, pp. 491-508, (1983).

20. Callow, G. D. and Peat, K. S., Insertion Loss of engine inflow and exhaust silencers, I Mech. E C19/88:39-46, (1988).

21. Hota, R. N. and Munjal, M. L., Approximate empirical expressions for the aeroacoustic source strength level of the exhaust system of compression ignition engines, International Journal of Aeroacoustics, 7(3&4), pp. 349-371, (2008).

22. Fairbrother, R., Boden, H. and Glav, R., Linear Acoustic Exhaust System Simulation Using Source Data from Linear Simulation, SAE Technical Paper series, 2005-01-2358, (2005).

23. Boden, H., Tonse, M. and Fairbrother, R., On extraction of IC-engine acoustic source data from non-linear simulations, Proceedings of the Eleventh International Congress on Sound and Vibration, (ICSV11), St. Petersburg, Russia, (2004).

24. Knutsson, M. and Boden, H., IC-Engine intake source data from non-linear simulations, SAE Technical Paper series, 2007-01-2209, (2007).

25. Hota, R. N. and Munjal, M. L., Intake source characterization of a compression ignition engine: empirical expressions, Noise Control Engineering Journal, 56(2), pp. 92-106, (2008).

26. Munjal, M. L. and Hota, R. N., Acoustic source characteristics of the exhaust and intake systems of a spark ignition engine, Paper No. 78, Inter-Noise 2010, Lisbon, Portugal, (2010).

27. BOOST Version 5.0.2, AVL LIST GmbH, Graz, Austria, (2007).

28. Munjal, M. L., Panigrahi, S. N. and Hota, R. N., FRITAmuff: A comprehensive platform for prediction of unmuffled and muffled exhaust noise of I.C. engines, 14th International Congress on Sound and Vibration, (ICSV14), Cairns, Australia, (2007).

29. Joint Departments of the Army, Air force and Navy, USA, Power Plant Acoustics, Technical manual TM 5-805-9 AFT 88-20 NAVFAC Dhf 3.14, (1983).

30. Bies, D. A. and Hansen, C. H., Engineering Noise Control: Theory and Practice, Fourth Edition, Spon Press, London, (2009).

31. Baumann, H. D. and Coney, W. B., Noise of Gas Flows, Chap. 15 in I. L. Ver and L. L. Beranek (Ed.), Noise and Vibration Control, Second Edition, John Wiley, New York, (2006).

32. International Standard Industrial-process control valves – Part 8-3: Noise considerations – Control valve aerodynamic noise prediction method, IEC 60534-3-3 Edition 3.0, (2010-11).

33. Cummings, A., Chang, I.-J. and Astley, R. J., Sound transmission at low frequencies through the walls of distorted circular ducts, Journal of Sound and Vibration, 97, pp. 261-286, (1984).

34. Munjal, M. L., Gowtham, G. S. H., Venkatesham, B. and Harikrishna Reddy, H., Prediction of breakout noise from an elliptical duct of finite length, Noise Control Engineering Journal, 58(3), pp. 319-327, (2010).

35. Venkatesham, B., Pathak, A. G. and Munjal, M. L., A one-dimensional model for prediction of breakout noise from a finite rectangular duct with different acoustic

boundary conditions, International Journal of Acoustics and Vibration, 12(3), pp. 91-98, (2007).

36. Narayana, T. S. S. and Munjal, M. L., Computational Prediction and measurement of break-out noise of mufflers, SAE Conference, SIAT 2007, SAE Paper 2007-26-040, ARAI, Pune, India, pp. 501-508, (2007).

37. Munjal, M. L. and Eriksson, L. J., An analytical, one-dimensional standing wave model of a linear active noise control system in a duct, Journal of the Acoustical Society of America, 84(3), pp. 1086-1093, (September 1988).

38. Munjal, M. L. and Eriksson, L. J., Analysis of a linear one-dimensional noise control system by means of block diagrams and transfer functions, Journal of Sound and Vibration, 129(3), pp. 443-455, (March 1989).

39. Hansen, C. H. and Snyder, S. D., Active Control of Sound and Vibration, London: E&FN Spon, (1997).

40. Hansen, C. H., Understanding Active Noise Cancellation. London: E&FN Spon, (2001).

41. Nelson, P. A. and Elliott, S. J., Active Control of Sound, London: Academic Press, (1992).

## Problems in Chapter 5

**Problem 5.1** Design a simple expansion chamber muffler for a TL peak of 15 dB at a frequency of 125 Hz, for plane waves in stationary medium at $25^0$C.

[**Ans.: Chamber length, $l$ = 0.692 m; area ratio, $m$ = 11.16**]

**Problem 5.2** Design a simple expansion chamber muffler for TL of at least 10 dB from 50 to 100 Hz for stationary medium at $25^0$C.

[**Ans.: Chamber length, $l$ = 1.153 m; area ratio, $m$ = 7.07**]

**Problem 5.3** Design a circular, double-tuned extended-tube expansion chamber (find out the extended inlet and outlet lengths, $l_a$ and $l_b$) of length 0.5 m, area ratio 9.0, inlet/outlet tube diameter and thickness equal to 40 mm and 1 mm, respectively, for stationary medium at $25^0$C.

[**Ans.: $l_{g,a}$ = 241.5 mm and $l_{g,b}$ = 116.5 mm** ]

**Problem 5.4** Design a tuned extended-tube concentric tube resonator within an axi-symmetric chamber with $l = 0.5$ m, $d = 40$ mm, $D = 120$ mm, perforate porosity, $\sigma = 0.1$.

$$[\text{Ans.}: l_{g,a} = 232.4 \text{ mm and } l_{g,b} = 107.4 \text{ mm }]$$

**Problem 5.5** A 2-m long, 800 mm×800 mm square duct with rigid walls is proposed to be lined on all four sides (see Fig. (a) below) with 100 mm thick mineral wool blanket with sound absorption coefficient of 0.6 in the 250 Hz octave band. Evaluate its TL. If the lining were re-configured so as to form a 4-pass parallel baffle muffler, as shown in Fig. (b), what would its TL?

Assume that the sound absorption coefficient of the 50-mm thick blanket is 0.3 in the 250-Hz band.

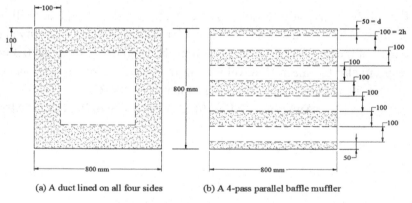

(a) A duct lined on all four sides     (b) A 4-pass parallel baffle muffler

$$[\text{Ans.}: 12 \text{ dB}; 18 \text{ dB}]$$

**Problem 5.6** Design a muffler (estimate the chamber volume and length of the 6mm diameter tail pipe) for a single-cylinder 50cc four-stroke-cycle engine running at 6000 RPM so as to ensure insertion loss of 8 dB at the firing frequency. Assume speed of sound in the exhaust gases to be 500 m/s.

$$[\text{Ans.: Chamber volume, } V_c = 500 \text{ cc; length of the tail pipe,}$$
$$l_t = 0.503 \text{ m; Alternatively } V_c = 1 \text{ liter and } l_t = 252 \text{ mm}]$$

**Problem 5.7**  Evaluate length of the 2.4 mm-diameter neck of a Helmholtz resonator opening into a cavity of 1 liter volume for the intake system of a 4-cylinder four-stroke diesel engine in order to suppress the firing frequency tone of the engine idling at 900 RPM. State the assumptions made in the process.

[**Ans.: Neck length,** $l_n$ = **11.95 mm**]

**Problem 5.8**  Evaluate and compare the back pressure of the following four muffler configurations:

(a) simple expansion chamber (Fig. 5.38)
(b) concentric tube resonator (Fig. 5.10)
(c) cross-flow, three-duct closed-end element (Fig. 5.13)
(d) plug muffler (Fig. 5.44)

For all the four configurations, diameter of the exhaust pipe as well as tail pipe is 40 mm; diameter of the circular shell is 160 mm; porosity and length of the perforates (in configurations b, c and d) are 0.05 and 0.3 m, respectively; the mean flow velocity in the exhaust pipe (as well as tail pipe) is 50 m/s and temperature of the medium is 500°C.

[**Ans.: (a) 770 Pa, (b) 51.4 Pa, (c) 2646 Pa, (d) 2679 Pa**]

Chapter 6

# Noise Control Strategies

Noise may be controlled at the source, in the path or at the receiver end. The cost of noise control increases as one moves away from the source. In fact, designing for quietness is the most cost effective way; prevention is better than cure! In this chapter, we list and illustrate all possible techniques and strategies for noise control of an existing machine as well as choice or design of quieter machines, processes and industrial layouts.

## 6.1 Control of Noise at the Source

Noise of a machine may be reduced at the source by adopting some of the following measures or practices.

### 6.1.1 Select a quieter machine

Select a quieter machine from the market even if it is relatively costlier. Often, this additional cost is less than the cost and hassle of the retrofit noise control measures. A machine would generally be quieter if its moving parts were fabricated to closer tolerances. Manufacturing mating parts to closer tolerances reduces micro-impacting between the parts. This in turn

- reduces the mechanical (impact) noise;
- reduces vibrations;
- reduces wear and tear;
- increases fatigue life;
- increases the interval between maintenance outings; and
- increases the accuracy of a machine tool.

Overall, manufacturing to closer tolerances is a good engineering practice and makes economic sense in the long run. In other words, the lifetime cost of a quieter, though costlier, machine is relatively lower. Human desire to obtain large power from small power packs, particularly for aeroengines, motorbikes and automobiles, has resulted in a race for high speed engines and turbines. However, noise control at the source can be achieved by selecting large, slow machines rather than small, faster ones, particularly for stationary installations of captive diesel generators, compressors, etc.

### 6.1.2  Select lossy materials

Mechanical impacting sets different parts of a machine into free vibration. The sheet metal components or covers, vibrating in their natural modes, radiate noise into the atmosphere. Free vibration of sheet metal components can be reduced by fabricating these components out of lossy materials like plastics; that is, materials with relatively high loss factor. We could consider replacing metal gears with plastic gears. Of course, such materials are very low in strength, and therefore, cannot be used in high-stress locations.  Alternatively, we could make use of free layer damping (FLD), or preferably, constrained layer damping (CLD) (discussed in Chapter 3 of this book) for the sheet metal components or covers exposed to the atmosphere. For example, circular saw blades should be replaced with damped blades.

### 6.1.3  Use quieter processes or tools

One very popular way of reducing noise in energy conversion is use of electrical motors rather than reciprocating engines or gas turbines.

Use of high velocity jets for cleaning must be avoided as far as possible in industry. We should seriously consider replacing pneumatic tools with electric tools, and pneumatic ejectors with mechanical ejectors.

Forging down to the net shape in a single step die is another source of high noise in workshops. We must consider replacing single operation dies with stepped dies, and rolling or forging with pressing.

Increasing the contact time during impacting reduces the rate of change of force. Therefore, rotating shears produce less noise than mechanical presses, which in turn are generally quieter than hammers. Welding or squeeze riveting creates less noise than impact riveting.

Increasing the contact area is a potent method for reducing noise at the source in material handling. Therefore, use of belt conveyers instead of roller conveyers recommends itself. For the same reason, helical gears have been observed to produce much less noise than spur gears, and spiral cutters are quieter than straight-edge cutters by upto 10 dB. Probably, the best example of noise reduction by increasing the contact area is use of belt drives in place of gear drives, and pneumatic tyres instead of solid steel wheels. As a matter of fact, development of pneumatic tyres made of rubber and nylon strings was probably the most important step towards reducing automobile noise and preserving the pavement.

### 6.1.4 *Reduce radiation efficiency*

For a given vibration amplitude, noise radiation from a structure depends upon radiation efficiency, which in turn depends on the shape and texture of the vibrating surfaces. Flat continuous plates are efficient radiators of sound, and therefore must be replaced with curved or ribbed panels, perforated plates, or woven strips of metal. Two sides of a plate radiate out-of-phase waves, and perforations expose both sides of the plate simultaneously, resulting in substantial cancellations at the holes or periodic gaps. This effect roughly represents conversion of monopole mechanism of noise generation into dipole mechanism, which has intrinsically much lower radiation efficiency.

This mechanism of noise reduction can be, and often is, utilized in design of hoppers, stillages or tote boxes for material handling operations. Often, the noise in a factory is increased dramatically when the metal ore, castings or waste material is tipped into a tote box made of sheet metal. If this totebox or stillage were made of interwoven metal strips, noise reduction of the order of 20 dB could be obtained. This represents a tremendous noise reduction at the source at minimal cost. This principle can also be applied to cover transmission couplings, baggage trolleys on railway platforms, side covers of mechanical hammers and presses, etc.

### 6.1.5 *Maintain for quietness*

Often, it is observed that a machine that has been in use for a while is no longer as quiet as it was when it was first installed. In order to avoid this additional noise at the source, it is necessary to

(a) have the rotating parts of the machine balanced on site, not at the supplier's premises;
(b) monitor the condition of the bearings continuously and have them lubricated regularly;
(c) replace or adjust the worn or loose parts as soon as detected; and
(d) follow the periodic maintenance schedules specified by the supplier.

If a machine is thus maintained, its noise level would not unduly increase with use or age. As indicated elsewhere in this book, increased noise is often a symptom of poor maintenance of the machine. If not attended to, the primary function of the machine would be compromised and chances of its failure or malfunction would increase alarmingly.

### 6.1.6 *Design of flow machinery for quietness*

Most often, noise is radiated by cooling equipments. As a rough estimate, half of all noise in the world emanates from fans, blowers, and heating ventilating and air-conditioning (HVAC) systems. Noise control of flow machinery at the source may be effected as follows.

(a) Provision of adequate cooling fins would reduce or eliminate the need of fan for forced convection cooling. Then, either the fan would not be needed at all, or a small, relatively quieter, fan would do.

(b) As will become clear later in this chapter, centrifugal fans are generally quieter than propeller fans of the same capacity working against the same back pressure. Again, backward curved blades produce less noise than the forward curved blades. Computational Fluid Dynamics, (CFD) can be used to design quieter blade profiles like airfoil blades.

(c) If the flow entering a fan does not have uniform velocity profile, or is highly turbulent, then the fan would produce more noise than usual. Therefore, we should locate the fans in smooth, undisturbed air flow, not too near a bend, for example.

(d) Large low speed fans are generally quieter than smaller faster ones.

(e) For shop floor cooling, it is better to provide each worker with small individual fans than installing large high speed fans at one end of the workshop.

(f) Noise in high velocity flow ducts increases substantially at the sharp bends due to separation of boundary layer. Therefore, adequate CFD modeling may be done and suitable guide vanes may be provided to avoid flow separation.

(g) Flow noise or aerodynamic noise from subsonic jets increases with sixth to eighth power of flow velocity. Therefore, we should design the flow ducts so as to maximize the cross-section of flow streams. Incidentally, this consideration led to the development and use of high bypass turbofan engines for transport aircraft. It reduced the jet noise by 10-20 dB while retaining the same thrust. Moreover, propulsion efficiency of turbofan engines is higher than the corresponding turbojet engines.

(h) Drastic reduction of pressure (as in high pressure venting of air or superheated steam) in a single step (single stage expansion trim valve) produces very high sound pressure levels. As explained before in Chapter 5, this may be reduced by means of a multistage expansion trim valve or a series of adequately designed orifice plates.

(i) As will be learnt later in this chapter, noise of a fan or blower increases with flow rate as well as the back pressure against which the flow needs to be pushed. Therefore, the flow passages need to designed so as to minimize the stagnation pressure drop in the system. Incidentally, it will also reduce the load on the fan and hence the power consumption.

(j)    Often, the nozzles used in high pressure pneumatic cleaning devices produce high jet noise. This can be reduced by means of specially designed multijet nozzles or by-pass nozzles.

## 6.2    Control of Noise in the Path

For any existing noisy machine, the option of noise control at the source as well as designing for quietness is not available. The next option is to control the noise in the path making use of acoustic barrier, hood, enclosure, etc. for the casing radiated noise, muffler or silencer for the duct borne intake/exhaust noise, vibration isolator to reduce propagation of the unbalanced forces to the foundation or the support structure, and structural discontinuities (impedance mismatch) to block propagation of the structure borne sound. All these measures have been discussed at length in Chapters 3, 4 and 5. As indicated there, damper plays the same role in absorption of structure-borne sound as acoustically absorptive material does in dissipation of airborne sound. This is illustrated hereunder.

|  | Reflective measures | Absorptive measures |
|---|---|---|
| Airborne sound | Sound barrier | Sound absorber |
| Structure-borne sound | Vibration isolator | Vibration damper |

In practice, however, noise control measures are not purely reflective (reactive) or purely absorptive (dissipative). For example, a stand-alone acoustic enclosure consists of both types of components — impervious layer for reflection and acoustic lining on the inside for absorption. Similarly, viscoelastic pads combine elasticity and damping.

Finally, it may be noted that in practice, despite the cost disadvantage, noise is controlled more often in the path than at the source, because of logistic convenience and ready availability of acoustic enclosures, mufflers or silencers, vibration isolators and dampers.

## 6.3  Noise Control at the Receiver End

This option is the last resort, as it were. The receiver may be protected from excessive noise exposure by means of

(a)  ear plugs or muffs (which are really a path control measure)
(b)  rotation of duties (part-time removal from the noisy environment)
(c)  cabin for the operator, driver or foreman (which is again a path control measure)
(d)  control room for the supervisory personnel in noisy test cells.

Ear plugs and muffs are readily available in the market. They provide an insertion loss of 5 to 15 dB for the user, but then the user's functionality may be compromised. He/she may not be able to detect a malfunction in the machine, or may not be forewarned of the damages arising from a malfunction in a machine in the vicinity.

Design of acoustic cabins and control rooms is essentially similar to the design of acoustic enclosures discussed in Chapter 4.

It may be noted that noise control at the receiver end may protect a particular receiver from the noise, but would not help others in the vicinity. This comment would also apply to some of the path control measures. Therefore, the noise control at the source or designing for quietness is the most cost-effective way that would help everybody in the vicinity, not just the operator or driver of the machine.

## 6.4  Noise Control of an Existing Facility

If there is a complaint about environmental noise due to an existing facility like captive power station, compressed air facility, an HVAC system, or a workshop, we may proceed as follows.

(a)  Make SPL measurements so as to identify and rank order different sources of noise.
(b)  Calculate the extent of noise reduction required for the noisiest of the sources (machines or processes) in order to reduce the total noise in the neighborhood to the desired level.

(c)   Identify the paths of noise transmission — structure borne as well as airborne – from the major sources to the neighborhood, and plan the path control measures.

(d)   Estimate the cost-effectiveness of different alternatives.

(e)   Carry out detailed engineering and installation of the selected noise control measures.

(f)   Verify the effectiveness of the implemented measures, and work out refinements or corrections as necessary to meet the environmental noise limits.

For community noise, it is advisable that at worst, any facility should not increase background (or ambient) noise levels in a community by more than 5 dB (A) over existing levels without the facility, irrespective of what regulations may allow.

If complaints arise from the work place, then regulations should be satisfied, but to minimize hearing damage compensation claims, the goal of any noise control program should be to reach a level of no more than 85 dB (A).

It is important to point out that we should not wait for complaints to arise. Once a person files an official complaint, he/she would not be satisfied by normal noise reduction. A better or more desirable way is to carry out an environmental impact assessment (EIA) before a potentially noisy facility is installed. If the projected noise levels at the property line exceed the mandated limits (depending on the category of the locality or area), then either quieter machines may be ordered/specified, or path control measures may be specified for the relatively noisy machines. This is discussed in the following sections.

The EIA exercise needs prior knowledge of the total sound power level, the spectral content, and the relative locations of all the noisy machines and processes that would constitute the proposed facility. The next few sections list empirical expressions for sound power levels (or sound pressure level at 1 m distance) of some of the most common (and noisiest) machines and processes, gleaned from the literature [1, 2]. These expressions, incidentally, give us a glimpse of the parameters and considerations for selection or design of quieter machines and processes.

This is in fact the primary purpose of listing these expressions here in this chapter dealing with noise control strategies.

## 6.5 Estimation and Control of Compressor Noise

Sound power level generated within the exit piping of large centrifugal compressors $(>75\ kW)$ is given by [3, 1].

$$L_w = 20\ \log kW + 50\ \log U - 46\quad dB \tag{6.1}$$

where U is the impeller tip speed $(m/s)$, $30 < U < 230$, and

$kW$ is the power of the driver motor in kilowatts.
The frequency of maximum (peak) noise level is given by

$$f_p = 4.1\ U\quad (Hz) \tag{6.2}$$

The sound power level in the octave band containing $f_p$ is taken as 4.5 dB less than the overall sound power level. The octave-band frequency spectrum rolls off at the rate of 3 dB per octave above and below the band of maximum noise level.

Equation (6.1) indicates that sound power level of large centrifugal compressor increases with second power of the driver motor power (power index = 2) and fifth power (power index = 5) of the impeller tip speed. Thus, $L_w$ is a much stronger function of the impeller tip speed, $U$. It is obvious that if we want to reduce the compressor noise, we must design it for the lowest tip speed:

$$U = \pi D.RPM/60\quad (m/s) \tag{6.3}$$

where D is diameter of the impeller, and RPM denotes the impeller rotational speed in revolutions per minute.

**Example 6.1** Evaluate the overall A-weighted sound power level of a centrifugal compressor of 100 $kW$ power with impeller diameter of 0.9 $m$ turning at 3000 RPM.

**Solution**

Use of Eq. (6.3) yields the impeller tip speed, U:

$$U = \frac{\pi \times 0.9 \times 3000}{60} = 141.4 \ m/s$$

Use of Eq. (6.1) yields

$$L_w = 20 \ \log 100 + 50 \ \log 141.4 - 46 = 101.5 \ dB$$

Eq. (6.2) gives the peak frequency, $f_p$:

$$f_p = 4.1 \times 141.4 = 580 \ Hz$$

Obviously, $f_p$ falls in the *500-Hz* octave band. In this band, the sound power level would be 101.5 − 4.5 = 97 *dB*, and in the neighboring bands it will fall at the rate of 3 *dB* per octave. The rest of the calculations are shown in the table below.

| Octave band mid-frequency (Hz) | 63 | 125 | 250 | 500 | 1000 | 2000 | 4000 | 8000 |
|---|---|---|---|---|---|---|---|---|
| Band sound power level | 88 | 91 | 94 | 97 | 94 | 91 | 88 | 85 |
| A-weighting correction (from Table 1.2) | −26.2 | −16.1 | −8.6 | −3.2 | 0 | 1.2 | 1.0 | −1.1 |
| A-weighted power level | 61.8 | 74.9 | 85.4 | 93.8 | 94 | 92.2 | 89.0 | 83.9 |

Finally, the overall A-weighted sound power level is evaluated by means of Eq. (1.36), making use of the last row of the table above. Thus, overall A-weighted sound power level of the compressor is given by

$$L_{wA} = 10 \ \log \left( 10^{6.18} + 10^{7.49} + 10^{8.54} + 10^{9.38} + 10^{9.4} + 10^{9.22} + 10^{8.9} + 10^{8.39} \right)$$
$$= 99.0 \ dB$$

For comparison, the overall sound power level within the exit piping of a reciprocating compressor is given by

$$L_W = 106.5 + 10 \ \log (kW) \tag{6.4}$$

Its octave-band frequency spectrum may be calculated using the same procedure as for the centrifugal compressor above, except that the peak frequency for a reciprocating compressor is given by

$$f_p = B \times RPM / 60 \tag{6.5}$$

where B is the number of cylinders of the compressor. Interestingly, a 100 *kW* reciprocating compressor would generate a sound power level of 136.5 *dB* as per Eq. (6.4), which is 25 *dB* more than the corresponding centrifugal compressor of Example 6.1 above.

Exterior noise levels of large compressors are given by the following empirical expression [4]:

Centrifugal Compressors:

$$L_W(\text{casing}) = 79 + 10 \, \log(kW) \tag{6.6}$$

$$L_W(inlet) = 80 + 10 \, \log(kW) \tag{6.7}$$

Rotary and reciprocating compressors (including partially muffled inlets):

$$L_W(\text{casing}) = 90 + 10 \, \log(kW) \tag{6.8}$$

Table 6.1. Octave band correction for exterior noise levels radiated by compressors, (adopted from Ref. [4]).

| Octave band Centre Frequency (Hz) | Correction (dB) | | |
|---|---|---|---|
| | Rotary and reciprocating | Centrifugal, casing | Centrifugal, air inlet |
| 31.5 | 11 | 10 | 18 |
| 63 | 15 | 10 | 16 |
| 125 | 10 | 11 | 14 |
| 250 | 11 | 13 | 10 |
| 500 | 13 | 13 | 8 |
| 1000 | 10 | 11 | 6 |
| 2000 | 5 | 7 | 5 |
| 4000 | 8 | 8 | 10 |
| 8000 | 15 | 12 | 16 |

Equations (6.6) – (6.8) indicate clearly that in general rotary and reciprocating compressors are noisier than the corresponding centrifugal compressors by $90-(79 \oplus 80) = 90 - 82.5 = 7.5 \ dB$. Therefore, for quieter installations, we must prefer centrifugal compressors to the rotary and reciprocating compressors.

The corresponding overall A-weighted sound power levels may be evaluated by making use of Table 6.1 and the procedure outlined above in Example 6.1. The values listed in Table 6.1 are to be subtracted from the overall sound power level.

Incidentally, compressors with lower power $(< 75 \ kW)$ are not much quieter than those with larger power. An estimate of sound pressure level at 1 m distance from the body of the compressor may be had from table 6.2, which lists the SPL values measured in early eighties; i.e., about three decades ago. Noise control technologies have evolved a lot since then and the SPL values of the present day compressors are known to be lower by several decibels [1]. In other words, the values listed in Table 6.2 are conservative (on the higher side) by 5 to 10 dB.

Table 6.2. Estimated sound pressure levels of small air compressors at 1 m distance in dB, (adopted from Ref. [2]).

| Octave band Centre Frequency (Hz) | Air compressor power (kW) | | |
|---|---|---|---|
| | Up to 1.5 | 2–6 | 7–75 |
| 31.5 | 82 | 87 | 92 |
| 63 | 81 | 84 | 87 |
| 125 | 81 | 84 | 87 |
| 250 | 80 | 83 | 86 |
| 500 | 83 | 86 | 89 |
| 1000 | 86 | 89 | 92 |
| 2000 | 86 | 89 | 92 |
| 4000 | 84 | 87 | 90 |
| 8000 | 81 | 84 | 87 |

It may be noted that SPL of compressors is a weak function of the air compressor power, $kW$.

The overall sound power level of a compressor may be evaluated from the overall sound pressure level listed in Table 6.2 by means of the Survey method:

$$SWL \equiv L_w = L_p \left(1 \ m\right) + 10 \ \log S_m \qquad (6.9)$$

where $S_m$ is area of the hypothetical rectangular surface surrounding the compressor, at *1 m* distance from the nominal compressor body.

Main sources of noise small reciprocating hermetically sealed piston compressors are [5]

(a)  gas flow pulsation through the inlet and discharge valves and pipes,
(b)  gas flow fluctuations in the shell cavity, which excite the cavity and shell modes,
(c)  vibrations caused by the mechanical system rotation of the drive shaft and out-of-balance reciprocating motion of the piston and connecting rod, and
(d)  electric motor noise.

All these mechanisms contribute directly or indirectly to the compressor shell vibration response and result in shell sound radiation. Therefore, noise control measures for these refrigerator compressors include

(i)  improved design of suction muffler
(ii)  vibro-acoustic design of the shell making use of FEM and/or BEM.

For medium and large compressors, noise control is normally done by means of intake mufflers and acoustic wrapping, hood or a stand-alone enclosure as discussed later in this chapter in the section 6.10.1 on turbine noise control. Designing the compressors of different types for quietness calls for a detailed knowledge of the working of the compressor, its constituent components and the interaction thereof [5].

## 6.6  Estimation and Control of Noise of Fans and Blowers

Sound power of axial fans and blowers (centrifugal fans) may be described in terms of the Specific Sound Power, H, defined by Madison [6] as follows:

$$H = \frac{W}{\left(\Delta P\right)^2 Q} \Rightarrow L_W = L_H + 10 \ \log Q + 20 \ \log\left(\Delta P\right) \qquad (6.10)$$

where $\Delta P$ is the static pressure rise, or the stagnation pressure drop or back pressure on the fan or blower, in Pascals, and Q is the volumetric flow rate in $m^3 / s$. Based on the research of Graham and Hoover [7] and others, Table 6.3 gives specific sound power levels ($L_H$) in the eight primary octave bands for different types of fans and blowers [8]. In the logarithmic form of Eq. (6.10), $\Delta P$ is in $kPa$ and the constant term arising from logarithms has been absorbed in $L_H$.

Table 6.3 Specific power levels in eight lowest octave bands for a variety of axial and centrifugal fans (adopted with permission from Ref. [8]).

| Fan Type | Rotor Diameter (m) | Octave Band Center Frequency (Hz) | | | | | | | | Add for BPF |
|---|---|---|---|---|---|---|---|---|---|---|
| | | 63 | 125 | 250 | 500 | 1k | 2k | 4k | 8k | |
| Backward-curved | >0.75 | 85 | 85 | 84 | 79 | 75 | 68 | 64 | 62 | 3 |
| centrifugal | <0.75 | 90 | 90 | 88 | 84 | 79 | 73 | 69 | 64 | 3 |
| Forward-curved centrifugal | All | 98 | 98 | 88 | 81 | 81 | 76 | 71 | 66 | 2 |
| Low-pressure radial | >1.0 | 101 | 92 | 88 | 84 | 82 | 77 | 74 | 71 | 7 |
| $996 \le \Delta P \le 2490$ | <1.0 | 112 | 104 | 98 | 88 | 87 | 84 | 79 | 76 | 7 |
| Mid-pressure radial | >1.0 | 103 | 99 | 90 | 87 | 83 | 78 | 74 | 71 | 8 |
| $2490 \le \Delta P \le 4982$ | <1.0 | 113 | 108 | 96 | 93 | 91 | 86 | 82 | 79 | 8 |
| High-pressure radial | >1.0 | 106 | 103 | 98 | 93 | 91 | 89 | 86 | 83 | 8 |
| $4982 \le \Delta P \le 14945$ | <1.0 | 116 | 112 | 104 | 99 | 99 | 97 | 94 | 91 | 8 |
| Vaneaxial | | | | | | | | | | |
| $0.3 \le D_h/D \le 0.4$ | All | 94 | 88 | 88 | 93 | 92 | 90 | 83 | 79 | 6 |
| $0.4 \le D_h/D \le 0.6$ | All | 94 | 88 | 91 | 88 | 86 | 81 | 75 | 73 | 6 |
| $0.6 \le D_h/D \le 0.8$ | All | 98 | 97 | 96 | 96 | 94 | 92 | 88 | 85 | 6 |
| Tubeaxial | >1.0 | 96 | 91 | 92 | 94 | 92 | 91 | 84 | 82 | 7 |
| | <1.0 | 93 | 92 | 94 | 98 | 97 | 96 | 88 | 85 | 7 |
| Propeller | All | 93 | 96 | 103 | 101 | 100 | 97 | 91 | 87 | 5 |

Here, BPF is the blade passing frequency, and the last column in Table 6.3 gives the value to be added to the level of the particular band in which the BPF falls.

**Example 6.2** Estimate the A-weighted sound power level of a 16-bladed backward-curved centrifugal fan running at 3000 *RPM* and delivering 50000 $m^3$/*hour* of air against back pressure of 5000 *Pa*.

**Solution**

The rotor tip diameter is *0.9 m*.

Flow rate $Q = \dfrac{50000}{3600} = 13.89 \ m^3 / s$

Back pressure, $\Delta P = \dfrac{5000}{1000} = 5 \ kPa$

$$10 \ \log Q + 20 \ \log (\Delta P) = 10 \ \log 13.89 + 20 \ \log 5$$
$$= 11.4 + 14.0 = 25.4 \ dB$$

$BPF = \dfrac{16 \times 3000}{60} = 800$ Hz, and this falls in the 1000-Hz octave band

So, referring to the first row of Table 6.3 we must add *3 dB* to the $L_H$ value of 75 dB at the 1000 Hz octave band. Thus, we can construct the following table

| Octave band centre frequency (Hz) | 63 | 125 | 250 | 500 | 1000 | 2000 | 4000 | 8000 |
|---|---|---|---|---|---|---|---|---|
| Specific sound power level, $L_H$ (dB) | 85 | 85 | 84 | 79 | 78 | 68 | 64 | 62 |
| $L_W$ (band) = $L_H$ + 25.4 (dB) | 110.4 | 110.4 | 109.4 | 104.4 | 103.4 | 93.4 | 89.4 | 87.4 |
| A-weighting correction | − 26.2 | − 16.1 | − 8.6 | − 3.2 | 0 | 1.2 | 1.0 | − 1.1 |
| $L_{WA}$ (octave) (dB) | 84.2 | 94.3 | 100.8 | 101.2 | 103.4 | 94.6 | 90.4 | 86.3 |

Finally, overall $L_{WA}$ may be obtained by logarithmically adding the octave band values of $L_{WA}$ from the last row. Thus, overall A-weighted SWL is given by

$$L_{WA} = 10 \, \log\left(\frac{10^{8.42} + 10^{9.43} + 10^{10.08} + 10^{10.12}}{+10^{10.34} + 10^{9.46} + 10^{9.04} + 10^{8.63}}\right) = 107.4 \, dB$$

Design of quieter fans (for automotive engine, for example) would normally involve

  (i)   reducing the fan impeller tip speed by reducing the RPM and/or impeller diameter,

  (ii)  increasing the number of blades so as to reduce the required RPM for a given flow rate and static pressure rise,

 (iii)  using thermal (temperature controlled) drive for the automobile fans so that the fan would switch off automatically at higher automobile speeds,

 (iv)  using airfoil blades in order to increase the aerodynamic efficiency,

  (v)  using non-metallic blades (for increased loss factor),

 (vi)  increasing the cooling efficiency so that a fan with lower tip speed would suffice,

 (vii)  minimizing restriction to the airflow so that the fan could work under lower back pressure,

(viii)  increasing the radiator frontal area and improving the radiator design in order to reduce the cooling load on the fan, and

 (ix)  using a shroud with minimal radial blade tip clearance in order to reduce the recirculation of flow around the tip.

For a given fan installation, noise can be reduced by means of

(a)  intake silencers

(b)  discharge or exhaust silencers

(c)  acoustic wrapping, hood or enclosure around the fan in order to contain and absorb the casing noise (discussed in Chapter 4).

A stand-alone acoustic enclosure would incorporate all three of these measures. The intake and exhaust silencers would then take the shape of parallel baffle mufflers or acoustic louvres (discussed in Chapter 5).

## 6.7 Estimation and Control of Noise of Packaged Chillers

Compressor is the main source of noise in packaged chillers. Measured values of SPL at *1 m* from the body of the chiller (or, for that matter, the compressor within) are listed in Table 6.4. These levels are generally higher than observed [1]. For more reliable values, we should refer to the manufacturer's data, if provided.

The octave band levels given in Table 6.4 may be first corrected for A-weighting and then added logarithmically to estimate the overall A-weighted sound pressure level of the packaged chiller. The chiller noise may be controlled by means of acoustic wrapping, hood or a stand-alone acoustic enclosure.

Table 6.4. Estimated sound pressure levels of packaged chillers at one meter, (adopted from Ref. [2]).

| Type and Cooling Capacity of Machine | Octave Band Center Frequency (Hz) | | | | | | | | | A-weighted (dBA) |
|---|---|---|---|---|---|---|---|---|---|---|
| | 31.5 | 63 | 125 | 250 | 500 | 1k | 2k | 4k | 8k | |
| Reciprocating compressors | | | | | | | | | | |
| 10 – 50 Tons | 79 | 83 | 84 | 85 | 86 | 84 | 82 | 78 | 72 | 89 |
| 51 – 200 Tons | 81 | 86 | 87 | 90 | 91 | 90 | 87 | 83 | 78 | 94 |
| Rotary screw compressors | | | | | | | | | | |
| 100 – 300 Tons | 70 | 76 | 80 | 92 | 89 | 85 | 80 | 75 | 73 | 90 |
| Centrifugal compressors | | | | | | | | | | |
| Under 500 Tons | 92 | 93 | 94 | 95 | 91 | 91 | 91 | 87 | 80 | 97 |
| 500 Tons or more | 92 | 93 | 94 | 95 | 93 | 98 | 98 | 93 | 87 | 103 |

## 6.8 Estimation and Control of Noise of Cooling Towers

Fan or blower is the main source of noise in the cooling towers, also known as remote radiators. Their sound power level, therefore, depends on the fan or blower used therein, and is given by the following empirical expressions [1, 2]:

Propeller-type cooling towers:

Fan power up to 75 kW:

$$L_W = 100 + 8 \, \log(kW) \tag{6.11}$$

Fan power greater than 75 kW:

$$L_W = 96 + 10 \, \log(kW) \qquad (6.12)$$

Centrifugal type cooling towers:
Fan power up to 60 kW:

$$L_W = 85 + 11 \, \log(kW) \qquad (6.13)$$

Fan power greater than 60 kW:

$$L_W = 93 + 7 \, \log(kW) \qquad (6.14)$$

The octave band sound power levels in different octave bands may be calculated by subtracting the values listed in Table 6.5, and finally, the overall A-weighted SWL can be calculated as per Example 6.2.

Equations (6.11) – (6.14) are functions of power only. It is understood that the fan or blower is running at its rated speed (RPM). If the speed is half of the rated speed, then SWL would be lower by about 8 dB [1]. Table 6.5 may be used to evaluate the sound power level in different octave bands. Finally, the overall A-weighted sound power level may be calculated as illustrated in Example 6.2 in Section 6.6 above.

Table 6.5. Values (dB) to be subtracted from the overall SWL of cooling towers to obtain the octave band SWL (adopted from Ref. [2]).

| Octave band centre Frequency (Hz) | Propeller type | Centrifugal type |
|---|---|---|
| 31.5 | 8 | 6 |
| 63 | 5 | 6 |
| 125 | 5 | 8 |
| 250 | 8 | 10 |
| 500 | 11 | 11 |
| 1000 | 15 | 13 |
| 2000 | 18 | 12 |
| 4000 | 21 | 18 |
| 8000 | 29 | 25 |
| A-weighted (dBA) | 9 | 7 |

Noise of the cooling tower can be reduced by enclosing it with an acoustic enclosure with the two longer opposite sides made entirely of acoustic louvers with sufficient air passages to provide adequate air for the cooling operation.

## 6.9 Estimation and Control of Pump Noise

Hydraulic pumps are used in processing plants, climate control systems, hydraulic fluid power, clean water supply, etc. Empirical expressions of the overall SPL of pumps at 1 m distance are listed in Table 6.6 for different speed ranges, as a function of the nominal drive motor power.

Table 6.6. Overall sound pressure levels at *1 m* from the pump (adopted from Ref. [2])*.

| Speed range (rpm) | Nominal Drive Motor Power | |
|---|---|---|
| | Under 75 kW | Above 75 kW |
| 3000 – 3600 | 72 + 10 log (kW) | 86 + 3 log (kW) |
| 1600 – 1800 | 75 + 10 log (kW) | 89 + 3 log (kW) |
| 1000 – 1500 | 70 + 10 log (kW) | 84 + 3 log (kW) |
| 450 – 900 | 68 + 10 log (kW) | 82 + 3 log (kW) |

* Subtract 2 dB to obtain the corresponding A-weighted values.

The octave-band frequency adjustments for the pump SPL are listed below in Table 6.7.

Table 6.7. Frequency adjustments for pump sound pressure levels (adopted from Ref [2]).

| Octave band centre Frequency (Hz) | Values to be subtracted from overall sound pressure level (dB) |
|---|---|
| 31.5 | 13 |
| 63 | 12 |
| 125 | 11 |
| 250 | 9 |
| 500 | 9 |
| 1000 | 6 |
| 2000 | 9 |
| 4000 | 13 |
| 8000 | 19 |

Table 6.8. Prediction of the A-weighted sound power level generated by different pumps (adapted with permission from Cudina [9]).

| Type of Pump | A-weighted Sound Power Level, $\left(P \text{ in } kW,\ P_{ref} = 1\ kW\right)$ in $dB$ | Valid for Power Consumption $P$ |
|---|---|---|
| Centrifugal pumps (single stage) | $L_{WA} = 71 + 13.5\ \log\ P \pm 7.5$ | $4\ kW \leq P \leq 2000\ kW$ |
| Centrifugal pumps (multistage) | $L_{WA} = 83.5 + 8.5\ \log\ P \pm 7.5$ | $4\ kW \leq P \leq 20000\ kW$ |
| Axial-flow pumps | $L_{WA} = 78.5 + 10\ \log\ P \pm 10$ at $Q_{BEP}$ | $10\ kW \leq P \leq 1300\ kW$ |
| | $L_{WA} = 21.5 + 10\ \log\ P$ $\pm 57\ Q/Q_{BEP} \pm 8$ | $0.77 \leq Q/Q_{BEP} \leq 1.25$ |
| Multi-piston pumps (inline) | $L_{WA} = 78 + 10\ \log\ P \pm 6$ | $1\ kW \leq P \leq 1000\ kW$ |
| Diaphragm pumps | $L_{WA} = 78 + 9\ \log\ P \pm 6$ | $1\ kW \leq P \leq 100\ kW$ |
| Screw pumps | $L_{WA} = 78 + 11\ \log\ P \pm 6$ | $1\ kW \leq P \leq 100\ kW$ |
| Gear pumps | $L_{WA} = 78 + 11\ \log\ P \pm 3$ | $1\ kW \leq P \leq 100\ kW$ |
| Lobe pumps | $L_{WA} = 84 + 11\ \log\ P \pm 5$ | $1\ kW \leq P \leq 10\ kW$ |

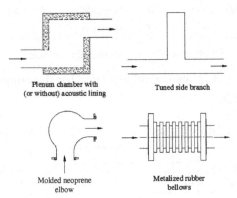

Plenum chamber with (or without) acoustic lining

Tuned side branch

Molded neoprene elbow

Metalized rubber bellows

Fig. 6.1 Schematics of devices for reduction of pressure pulsation in piping [9].

The overall A-weighted SPL at 1 m may be calculated as explained in Example 6.2 and SWL may be calculated by means of Eq. (6.10). Overall A-weighted sound power level of pumps of different types may directly be calculated by means of Table 6.8 (adapted from Cudina [9]).

Operation of a pump creates pressure pulsation which can be spread by pipes as structure-borne noise and by the liquid medium as fluid-borne noise around the whole pumping system. Figure 6.1 shows some devices for reduction of pressure pulsation in piping (structure-borne noise), and Figure 6.2 illustrates some measures for reduction of structural vibration [9].

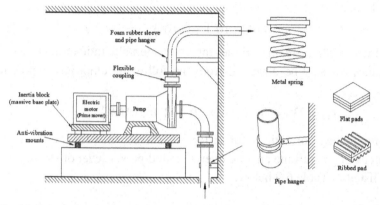

Fig. 6.2 Schematic of different measures for reduction of structure-borne sound [9].

## 6.10 Estimation and Control of Noise of Prime Movers

Conventional prime movers are turbines, reciprocating piston engines (particularly diesel engines and gas driven engines), and electric motors. Empirical expressions and the associated octave-band frequency corrections, all adopted from Refs. [1, 2], are listed below. As indicated earlier, the primary source of data [2] being nearly three decades old, the estimates are considerably on the higher side; the present-day prime movers are designed, fabricated and maintained to be relatively quieter by several decibels [1].

### 6.10.1 *Turbines*

Sources of noise in a typical turbine are the casing, inlet and exhaust. The overall sound power level of these sources for a gas turbine are given by the following empirical expressions [2], followed by some comments on the control measures:

Casing:

$$L_W = 120 + 5 \ \log(MW) \tag{6.15}$$

242                          *Noise and Vibration Control*

Inlet:

$$L_W = 127 + 5 \, \log(MW) \tag{6.16}$$

Exhaust:

$$L_W = 133 + 10 \, \log(MW) \tag{6.17}$$

Incidentally, steam turbines are considerably quieter than the gas turbines, as may be observed from the following empirical expressions [4]:

$$L_W = 93 + 4 \, \log(kW) = 105 + 4 \, \log(MW) \tag{6.18}$$

Frequency corrections for different octave bands for the three noise sources of gas turbine and the overall sound power level of steam turbine are listed in Table 6.9 below.

Table 6.9. Frequency adjustments (in dB) for gas turbine and steam turbine noise levels. Subtract these values from the overall sound power level $L_W$, to obtain octave band and A-weighted sound power levels (adopted from Refs. [1, 2]).

| Octave band centre frequency (Hz) | Value to be subtracted from overall $L_W$ (dB) | | | |
|---|---|---|---|---|
| | Gas Turbine | | | Steam turbine |
| | Casing | Inlet | Exhaust | |
| 31.5 | 10 | 19 | 12 | 11 |
| 63 | 7 | 18 | 8 | 7 |
| 125 | 5 | 17 | 6 | 6 |
| 250 | 4 | 17 | 6 | 9 |
| 500 | 4 | 14 | 7 | 10 |
| 1000 | 4 | 8 | 9 | 10 |
| 2000 | 4 | 3 | 11 | 12 |
| 4000 | 4 | 3 | 15 | 13 |
| 8000 | 4 | 6 | 21 | 17 |
| A-weighted (dB(A)) | 2 | 0 | 4 | 5 |

It may be observed from Eqs. (6.15 – 6.17) that exhaust noise is the largest source of turbine noise, followed by the intake noise and the casing noise in that order. The duct-borne exhaust noise and intake noise are absorbed by means of dissipative parallel baffle mufflers designed as per Chapter 5, and the casing noise is contained by means of acoustic lagging, hood or enclosure designed as per Chapter 4. Table 6.10 lists the noise reduction or insertion loss in the turbine casing noise due to different noise control treatments [1, 2].

Table 6.10. Approximate noise reduction of gas turbine casing enclosures (adopted from Refs. [1, 2]).

| Octave band centre frequency (Hz) | Type 1[a] | Type 2[b] | Type 3[c] | Type 4[d] | Type 5[e] |
|---|---|---|---|---|---|
| 31.5 | 2 | 4 | 1 | 3 | 6 |
| 63 | 2 | 5 | 1 | 4 | 7 |
| 125 | 2 | 5 | 1 | 4 | 8 |
| 250 | 3 | 6 | 2 | 5 | 9 |
| 500 | 3 | 6 | 2 | 6 | 10 |
| 1000 | 3 | 7 | 2 | 7 | 11 |
| 2000 | 4 | 8 | 2 | 8 | 12 |
| 4000 | 5 | 9 | 3 | 8 | 13 |
| 8000 | 6 | 10 | 3 | 8 | 14 |

a  Glass fiber or mineral wool thermal insulation with lightweight foil cover over the insulation

b  Glass fiber or mineral wool thermal insulation covered with a minimum 20 gauge aluminium or 24 gauge steel.

c  Enclosing metal cabinet for the entire packaged assembly, with open ventilation holes and with no acoustic absorptive lining inside the cabinet.

d  Enclosing metal cabinet for the entire packaged assembly, with open ventilation holes and with acoustic absorptive lining inside the cabinet.

e  Enclosing metal cabinet for the entire packaged assembly with all ventilation holes into the cabinet muffled and with acoustic absorptive lining inside the cabinet.

Unfortunately, however, Table 6.10 does not give the overall reduction in the A-weighted sound power Level due to each of the five treatments. This may be calculated as illustrated in Example 6.3 below.

**Example 6.3** Calculate the A-weighting corrections for the casing noise of a 2 *MW* gas turbine for each of the five acoustic treatments listed in Table 6.10.

**Solution**

Using Eq. (6.15) we get the overall sound power level generated by the casing of the 2-*MW* gas turbine as follows:

$$L_W = 120 + 5 \ \log(2) = 122 \ dB$$

Making use of the casing column of Table 6.9 and the octave-band noise reduction (NR) due to the five types of treatment from Table 6.10 helps us to construct the following table.

| Octave band mid-frequency (Hz) | 63 | 125 | 250 | 500 | 1k | 2k | 4k | 8k |
|---|---|---|---|---|---|---|---|---|
| (a) Value to be abstracted from $L_W$ (dB) | 7 | 5 | 4 | 4 | 4 | 4 | 4 | 4 |
| (b) Octave-band $L_W$ =121.5 – (a) (dB) | 115 | 117 | 118 | 118 | 118 | 118 | 118 | 118 |
| (c) A-weighted correction (dB) | – 26 | – 16 | – 9 | – 3 | 0 | 1 | 1 | – 1 |
| (d) Octave-band $L_{WA}$ = (b)+(c) (dB) | 88 | 100 | 109 | 114 | 118 | 119 | 119 | 116 |
| (e) NR due to treatment type 1 (dB) | 2 | 2 | 3 | 3 | 3 | 4 | 5 | 6 |
| (f) NR due to treatment type 2 (dB) | 5 | 5 | 6 | 6 | 7 | 8 | 9 | 10 |
| (g) NR due to treatment type 3 (dB) | 1 | 1 | 2 | 2 | 2 | 2 | 3 | 3 |
| (h) NR due to treatment type 4 (dB) | 4 | 4 | 5 | 6 | 7 | 8 | 8 | 8 |
| (i) NR due to treatment type 5 (dB) | 7 | 8 | 9 | 10 | 11 | 12 | 13 | 14 |
| (j) $L_{WA}$ with treatment type 1=(d)-(e) (dB) | 86 | 98 | 106 | 111 | 115 | 115 | 114 | 110 |
| (k) $L_{WA}$ with treatment type 2=(d)-(f) (dB) | 83 | 95 | 103 | 108 | 111 | 111 | 110 | 106 |
| (l) $L_{WA}$ with treatment type 3=(d)-(g) (dB) | 87 | 99 | 107 | 112 | 116 | 117 | 116 | 113 |
| (m) $L_{WA}$ with treatment type 4=(d)-(h) (dB) | 84 | 96 | 104 | 108 | 111 | 111 | 111 | 108 |
| (n) $L_{WA}$ with treatment type 5=(d)-(i) (dB) | 81 | 92 | 100 | 104 | 107 | 107 | 106 | 102 |

Logarithmic addition of the octave band values by means of Eq. (1.37) yields:

row (b) : $L_W$ without any treatment (bare casing) = 126.1 dB

row (d) : $L_{WA}$ without any treatment (bare casing) = 124.5 dB

The overall A-weighted correction for the bare casing = 124.5 – 126.1
= – 1.6 dB

row (j) : $L_{WA}$ with treatment type 1 = 120.4 dB ⇒ overall NR = 124.5 – 120.4 = 4.1 dBA

row (k) : $L_{WA}$ with treatment type 2 = 116.6 dB ⇒ overall NR = 124.5 – 116.6 = 7.9 dBA

row (l) : $L_{WA}$ with treatment type 3 = 122.1 dB ⇒ overall NR = 124.5 – 122.1 = 2.4 dBA

row (m) : $L_{WA}$ with treatment type 4 = 117.1 dB ⇒ overall NR = 124.5 – 117.1 = 7.4 dBA

row (n) : $L_{WA}$ with treatment type 5 = 112.6 dB ⇒ overall NR = 124.5 – 112.6. = 11.9 dBA

The following observations may be made from the results of Example 6.3:

(i)   For the noise radiated by the casing of a gas turbine, overall A-weighting correction is –1.6 dB. When rounded up to integer value, this tallies with the value of –2 dB (2 dB to be subtracted from the total linear sound power level) given in the last row of Table 6.9.

(ii)  Simply wrapping the turbine (or for that matter, any noisy machine) with a blanket of absorptive material gives a noise reduction of only 4.1 dBA (see row j in the example above); thin protective foil (of about 40 micron or 0.04 mm thickness) is of no consequence, acoustically speaking.

(iii) Replacing the thin foil (of negligible surface density) with 24 gauge steel (thickness about 0.8 mm, and surface density of $7800 \times 0.8 / 1000 = 6.24 \ kg/m^2$) helps considerably in that the noise reduction increases from 4.1 dBA to 7.9 dBA.

(iv)  Enclosing the turbine with a metal cabinet without any acoustic lining and keeping the ventilation holes open is more or less useless in that it reduces noise by 2.4 dBA only. This follows from the relationship between insertion loss (IL) and transmission loss (TL), as explained in Chapter 4.

(v)  Lining the metal cabinet with an acoustical blanket on the inside, while keeping the ventilation holes open, restores the insertion loss or noise reduction to 7.4 dBA, as for the type 2 treatment.

(vi) Finally, a proper acoustical cabinet with all ventilation holes or openings muffled yields a noise reduction of 11.9 dBA, or about 12 dBA. However this is much less than the insertion loss of 20 dBA that can be obtained by means of a properly designed, fabricated and installed stand-alone (or walk-around) acoustical enclosure, as described in Chapter 4. This brings out the inherent limitations of an acoustic hood or cabinet because of the dynamic or inertial coupling of the thin layer of air inbetween the machine (turbine, in this case) and tight fitting hood or cabinet.

### 6.10.2  Diesel engines

Diesel engines are used widely in automobiles as well as captive diesel generator (DG) sets. Empirical expression for the sound power level of the diesel engine exhaust noise is given by [1, 2]

$$L_W = 120 + 10 \, \log(kW) - K - (l_{ex}/1.2) \qquad (6.19)$$

where $K = 0$ for a naturally aspirated engine, and $K = 6$ for a turbocharged engine; $kW$ denotes the nominal rated power of the engine in kilowatts; and $l_{ex}$ is length of the long exhaust pipe (or tailpipe) in meters. Eq. (6.19) implies that attenuation due to wall friction and eddy losses is 0.83 dB per meter length of the exhaust pipe.

The octave band frequency spectrum of exhaust noise may be obtained by subtracting the values listed in Table 6.11 from SWL of Eq. (6.19).

Sound power level of the turbocharged diesel engine inlet noise is given by the empirical expression [2]

$$L_W = 95 + 5 \, \log(kW) - K - (l_{in}/1.8) \qquad (6.20)$$

where $l_{in}$ is length of the inlet or induction pipe. Eq. (6.20) implies that attenuation of the intake noise due to wall friction and eddy losses is about 0.56 dB per meter length of the induction pipe.

The octave band frequency spectrum may be obtained by means of Eq. (6.20) and Table 6.12.

Table 6.11. Frequency adjustments for unmuffled engine exhaust noise and A-weighted level (adopted from Ref. [2]).

| Octave band centre frequency (Hz) | Value to be subtracted from overall sound power level (dB) |
|---|---|
| 31.5 | 5 |
| 63 | 9 |
| 125 | 3 |
| 250 | 7 |
| 500 | 15 |
| 1000 | 19 |
| 2000 | 25 |
| 4000 | 35 |
| 8000 | 43 |
| A-weighted (dB(A)) | 12 |

Table 6.12. Frequency adjustments (dB) for turbo-charger air inlet. Subtract these values from the overall sound power level (Equation 6.20) to obtain octave band and A-weighted level (adopted from Ref. [2]).

| Octave band centre frequency (Hz) | Value to be subtracted from overall sound power level (dB) |
|---|---|
| 31.5 | 4 |
| 63 | 11 |
| 125 | 13 |
| 250 | 13 |
| 500 | 12 |
| 1000 | 9 |
| 2000 | 8 |
| 4000 | 9 |
| 8000 | 17 |
| A-weighted (dB(A)) | 3 |

Sound power level generated by the diesel engine casing is given by the empirical expression [1, 2]

$$L_W = 94 + 10 \ \log(kW) + A + B + C + D \qquad (6.21)$$

Constants A, B, C and D are listed below in Table 6.13. The octave band frequency spectrum of the diesel engine casing noise may be evaluated by subtracting the corrections of Table 6.14 from Eq. (6.21).

Table 6.13. Correction terms to be applied to Equation (6.21) for estimating the overall sound power level of the casing noise of a reciprocating engine (adopted from Ref. [2]).

| | |
|---|---:|
| Speed correction term, A | |
|     Under 600 rpm | − 5 |
|     600-1500 rpm | − 2 |
|     Above 1500 rpm | 0 |
| Fuel correction term, B | |
|     Diesel only | 0 |
|     Diesel and natural gas | 0 |
|     Natural gas only (including a small amount of pilot oil) | − 3 |
| Cylinder arrangement, C | |
|     In-line | 0 |
|     V-type | − 1 |
|     Radial | − 1 |
| Air intake correction term, D | |
|     Unducted air inlet to unmuffled roots blower | + 3 |
|     Ducted air from outside the enclosure | 0 |
|     Muffled roots blower | 0 |
|     All other inlets (with or without turbo-charger) | 0 |

### 6.10.3  *Electric motors*

A rough (and conservative) estimate of the sound pressure level (at 1 m) of the totally enclosed, fan cooled (TEFC) small electric motors is given by [1, 2]

Under 40 kW:

$$L_p(1m) = 17 + 17 \ \log(kW) + 15 \ \log(RPM) \qquad (6.22)$$

Over 40 kW (upto 300 kW):

$$L_p(1m) = 29 + 10 \ \log(kW) + 15 \ \log(RPM) \qquad (6.23)$$

The corresponding overall sound power level may be obtained by means of Eq. (6.9). Drip proof (DRPR) motors are known to be quieter than their TEFC counterparts by about 5 dB for the same power and RPM.

Table 6.14. Frequency adjustments (dB) for casing noise of reciprocating engines: subtract values from the overall sound power level (Equation (6.21)) to obtain octave band and A-weighted levels (adopted from Refs. [1, 2])

| Octave band centre frequency (Hz) | Engine speed under 600 rpm | Engine speed 600-1500 rpm | | Engine speed over 1500 rpm |
|---|---|---|---|---|
| | | Without roots blower | With roots blower | |
| 31.5 | 12 | 14 | 22 | 22 |
| 63 | 12 | 9 | 16 | 14 |
| 125 | 6 | 7 | 18 | 7 |
| 250 | 5 | 8 | 14 | 7 |
| 500 | 7 | 7 | 3 | 8 |
| 1000 | 9 | 7 | 4 | 6 |
| 2000 | 12 | 9 | 10 | 7 |
| 4000 | 18 | 13 | 15 | 13 |
| 8000 | 28 | 19 | 26 | 20 |
| A-weighted (dB(A) | 4 | 3 | 1 | 2 |

Table 6.15. Octave band level adjustments (dB) for small electric motors (adapted from Ref. [2]).

| Octave band centre Frequency (Hz) | Totally enclosed, fan cooled (TEFC) motor | Drip proof (DRPR) motor |
|---|---|---|
| 31.5 | 14 | 9 |
| 63 | 14 | 9 |
| 125 | 11 | 7 |
| 250 | 9 | 7 |
| 500 | 6 | 6 |
| 1000 | 6 | 9 |
| 2000 | 7 | 12 |
| 4000 | 12 | 18 |
| 8000 | 20 | 27 |
| A-weighted (dB(A)) | 1 | 4 |

Table 6.16. Sound power levels of large electric motors (adapted from Refs. [1, 2]).

| Octave band centre frequency (Hz) | 1800 and 3600 rpm | 1200 rpm | 900 rpm | 720 rpm and lower | 250 and 400 rpm vertical |
|---|---|---|---|---|---|
| 31.5 | 94 | 88 | 88 | 88 | 86 |
| 63 | 96 | 90 | 93 | 90 | 87 |
| 125 | 98 | 92 | 92 | 92 | 88 |
| 250 | 98 | 93 | 93 | 93 | 88 |
| 500 | 98 | 93 | 93 | 93 | 88 |
| 1000 | 98 | 93 | 96 | 98 | 98 |
| 2000 | 98 | 98 | 96 | 92 | 88 |
| 4000 | 95 | 88 | 88 | 83 | 78 |
| 8000 | 88 | 81 | 81 | 75 | 68 |

The octave band frequency adjustments, adapted from Refs. [1, 2] are listed in Table 6.15.

For large electric motors (rated power of 750 kW to 4000 kW) with nominal acoustical hoods, one can make use of Table 6.16 for a rough estimation of the sound power level, $L_w$ [1, 2].

In order to estimate SWL of motors with rated power above 4000 kW, we should add 3 dB to all levels in Table 6.15, and for motors rated between 300 to 750 kW, we should subtract 3 dB from all sound power levels. The overall linear sound power levels may be obtained by means of logarithmic addition. Similarly, the overall A-weighted sound power levels may be obtained by first adding algebraically the A-weighting corrections, and then adding up logarithmically, as illustrated before in the solved examples (6.1 – 6.3).

The sources of noise and vibration in electric motors (and indeed in most electric machines including generators or alternators) are [10]:

(a) electromagnetic forces in the airgap between the stator and rotor characterized by rotating or pulsating power waves in the range of 100 to 4000 Hz;

(b) bearings, depending on the quality of manufacture, the accuracy of machining of the bearing seats and the vibroacoustic properties of the end brackets;

(c) aerodynamic forces that depend on the construction of the fan and ventilation channels of the machine;

(d) mechanical imbalance of rotors that may excite an appreciable amount of vibration, particularly in high speed motors; and

(e) rubbing of brushes against the commutator or contact rings that produces a predominately high frequency noise.

The noise control of electric motors, therefore, calls for

(i) precision fabrication and installation of stator and rotor windings, rotors and bearings;

(ii) balancing of the rotor at site, at the time of installation, and periodic balancing thereafter; and

(iii) use of quieter (or muffled) cooling fans of TEFC motors.

## 6.11 Jet Noise Estimation and Control

Jet noise predominates in turbojet engines, furnaces, pneumatic cleaning devices, etc. The total acoustic power $(W_o)$ radiated by a high velocity jet is a small, but significant, fraction of the mechanical power $(W_m)$ of the jet. It is given by [1]

$$L_W = 10 \log W_a + 120, \quad W_a = \eta W_m, \quad W_m = \dot{m}U^2/2 = \rho_0 U^3 \pi d^2/8 \quad (6.24)$$

where   d is diameter (or equivalent diameter) of the jet,

$\dot{m}$  is the mass flux of the jet,

$U$  is the mean flow velocity averaged over the cross-section of the jet,

$\rho$  is mass density of the medium, and

$\eta$  is the acoustical efficiency given by the approximate expression [3]

$$\eta \approx \left(\frac{T}{T_0}\right)^2 \left(\frac{\rho}{\rho_0}\right) K_a M^5 \quad (6.25)$$

Here, $T_0$ is the absolute temperature $(in\,^oK)$ and $\rho_0$ is the mass density of the ambient medium; $T$ and $\rho$ are the absolute temperature and mass density of the medium of the jet; $M = U/c_0$ is Mach number relative to the ambient gas; and $K_a$ is the acoustical power coefficient given by

$$K_a \approx 5\times10^{-5} \text{ (for subsonic jets)} \qquad (6.26)$$

For choked (underexpanded) or supersonic jets, i.e., when pressure ratio exceeds 1.89 (for air medium), $\eta$ is given by

$$\eta = \eta_{\text{turb}} + \eta_{\text{shock}} \approx \eta_{\text{shock}} = \begin{cases} 3.16\times10^{-8}\ (PR)^{11.67}\ for\ PR \le 2.6 \\ 1.1\times10^{-3}\ (PR)^{0.745}\ for\ PR \ge 2.6 \end{cases} \qquad (6.27)$$

where PR denotes the pressure ratio of the jet

It may be noted that for a pressure ratio higher than the critical pressure ratio, $\eta$ may be much higher than (of the order of 100 times) the value given by Eq. (6.25). In other words, presence of shocks in the nozzle or just outside may generate sound power levels that could be 20 (or more) dB higher than the turbulence noise expressions (6.24 – 6.26).

The jet noise is highly directional, and this directivity for choked jets is different from that of the subsonic jets, as can be seen from Table 6.17.

Table 6.17. Directional correction for jets (adapted from Heitner [1, 3]).

| Angle from jet axis $(^o)$ | Directivity index, DI (dB) | |
|:---:|:---:|:---:|
| | Sub-sonic | Choked |
| 0 | 0 | – 3 |
| 20 | + 1 | + 1 |
| 40 | + 8 | + 6 |
| 60 | + 2 | + 3 |
| 80 | – 4 | – 1 |
| 100 | – 8 | – 1 |
| 120 | – 11 | – 4 |
| 140 | – 13 | – 6 |
| 160 | – 15 | – 8 |
| 180 | – 17 | – 10 |

Finally, the frequency spectrum may be evaluated by

$$L_p\left(f\right)-L_p(overall)=-5-16.045\left\{\log\left(f/f_p\right)\right\}^2+2.406\left\{\log\left(f/f_p\right)\right\}^4 \quad (dB)$$
(6.28)

where $f_p$ is the peak frequency given by the following approximate expression in terms of Strouhal number, $N_s$ [3]

$$N_s = f_p d/U, \quad N_s = 0.2 \text{ for subsonic jets} \quad (6.29)$$

This formula is an approximation inasmuch as it represents a curve that would be symmetric about $f = f_p$, but the curve in practice is not exactly symmetric [1, 3].

The pronounced directivity pattern of jets is of great significance for high vertical exhaust stacks or chimneys. Then, all personnel on ground are at an angle of more than $90^0$ with respect to the vertical jet axis. This is a significant noise control measure and should be duly accounted for in an EIA study.

Equations (6.24) and (6.25), when combined, indicate that the sound power generated by a fully expanded jet increases with the eighth power of the jet velocity. This underlines the necessity of reducing velocity of the jet by increasing the diameter of the jet (for the same mass flow or thrust), if we want to reduce the jet noise at the source. However, when this option is not available, or has already been utilized to the extent it was feasible, then we may make use of one of the four types of discharge silencers shown below in Fig. 6.3 [1].

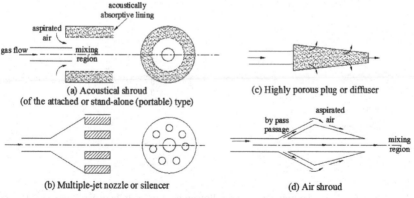

Fig. 6.3 Schematics of typical discharge silencers [1].

Noise from safety valves or relief valves with high pressure (much more than the critical pressure corresponding to sonic jet) may be reduced by means of a multi-stage trim (Fig. 6.4) or a set of properly designed multiple hole orifice plates (Fig. 6.5), adapted from Baumann and Coney [11].

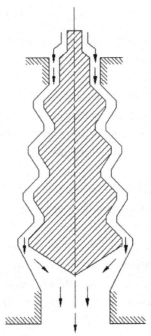

Fig. 6.4  Schematic of single flow multi-step valve plug [11].

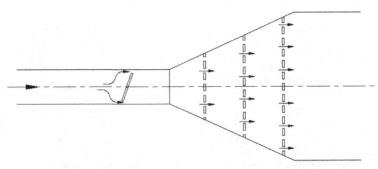

Fig. 6.5  Schematic of a butterfly valve followed by 3 multi-hole orifice plates for multi-stage pressure reduction [11].

**Example 6.4** 30 kg/s of air is being let out to atmosphere through a 11.5 m high vertical exhaust stack of 0.5 m diameter at 25°C. Estimate the SPL of the jet noise at a location 10 m away from the base of the stack, 1.5 m above the ground.

**Solution**

Density, $\rho_0 = \dfrac{p_0}{RT_0} = \dfrac{1.013 \times 10^5}{287 \times (273 + 25)} = 1.184 \ kg/m^3$

Sound speed, $c_0 = (\lambda RT)^{1/2} = (1.4 \times 287 \times 298)^{1/2} = 346.0 \ m/s$

Jet velocity, $U = \dfrac{\dot{m}}{\rho_0 A} = \dfrac{30}{1.184 \left( \dfrac{\pi}{4} \times (0.5)^2 \right)} = 129 \ m/s$

Mean flow Mach number, $M = \dfrac{U}{c_0} = \dfrac{129}{346} = 0.373$ (jet is subsonic)

Mechanical power of the jet,

$$W_m = \dfrac{\dot{m}U^2}{2} = \dfrac{30 \times (129)^2}{2} = 2.5 \times 10^5 \ W$$

Acoustical efficiency, $\eta = K_a M^5 = 5 \times 10^{-5} \times (0.373)^5 = 3.61 \times 10^{-7}$

Acoustical power, $W_a = \eta W_m = 3.61 \times 10^{-7} \times 2.5 \times 10^5 = 0.09 \ Watt$

Sound power level,

$$L_W = 10 \ \log \ W_a + 120 = 10 \ \log(0.09) + 120 = 109.5 \ dB$$

Distance of the receiver with respect to the exhaust point of the stack is 10 m horizontal and 11.5 − 1.5 = 10 m vertical. Thus, the receiver is located at an angle of 90 + 45 = 135° with respect to the jet. At this angle for a subsonic jet, referring to Table 6.17, directivity index, DI = − 12.5 dB

$$\therefore \ L_W \left( 135^0 \right) = L_W + DI = 109.5 - 12.5 = 97 \ dB$$

Distance of the observer from the jet exhaust,

$$R = \left( 10^2 + 10^2 \right)^{1/2} = 14.14 \ m$$

Finally, $L_p(R) = L_W - 10 \ \log\left(4\pi R^2\right)$

$$= 97 - 10 \ \log\left(4\pi \times 14.14^2\right)$$

$$= 97 - 34 = 63 \ \ dB$$

## 6.12 Estimation and Control of Gear Noise

Gears are used extensively in transmission of power in automobiles, earthmoving equipment, ship propulsion, captive power stations, home appliances, etc. Whine and rattle are two different types of gear noise. Whine is a tonal sound at the gear meshing frequency and gear rattle is an impulsive sound that occurs in lightly loaded gears excited by an externally applied oscillating torque [12].

The octave band sound pressure levels at all bands above 125 Hz, at a distance of 1 m from the spur gear system, may be estimated from the empirical expression [1, 2]

$$L_p(1 \ m) = 78 + 4 \ \log(kW) + 3 \ \log(RPM) \qquad (dB) \qquad (6.30)$$

where RPM is the rotational speed of the slowest gear shaft.

For estimation of SPL in the 63-Hz and 31.5 Hz octave bands, one may subtract 3 dB and 6 dB, respectively, from the SPL given by Eq. (6.30).

As indicated earlier in this chapter, helical gears are substantially quieter than spur gears for transmission of the same power. To be more precise, the actual noise reduction (compared to a straight spur gear) is given by the approximate relationship [1]

$$\Delta SPL = 13 + 20 \ \log Q_a \qquad (6.31)$$

where $Q_a$ is the number of teeth that would be intersected by a straight line parallel to the gear shaft. Thus, noise reduction of upto 30 dB may be obtained by means of well designed helical gears. Table 6.18 lists possible noise reductions consequent to change in certain design parameters [12].

It may, however, be noted that the many reductions that are possible and listed in Table 6.18 are not additive in nature.

In general, gear noise typically increases with load and power. More importantly, gear noise increases at the rate of 9-11 dB per octave beneath the torsional natural frequency [12]. Gear noise can be reduced by lowering roll angles, use of lead crown, avoiding profile errors and pressure angle errors, increasing lubricant viscosity, use of the split path drives and planetary gears, damping, use of bush bearings rather than rolling element bearings, and use of housing mounts and tuned absorbers [12].

Table 6.18. Effects of different gear design and manufacturing parameters on gear noise (adapted with permission from Houser [12]).

| | Direction to Reduce Noise | Noise Reduction (dB) | Comments |
|---|---|---|---|
| Number of teeth | Decrease | 0 – 6 | Lowers mesh frequency. |
| Contact ratio | Increase | 0 – 20 | Requires accurate lead and profile modifications |
| Helix angle | Increase | 0 – 20 | Machining errors have less effect with helical gears. Little improvement above about $35^0$. |
| Surface finish | Reduce | 0 – 7 | Depends on initial finish – reduces friction excitation. |
| Profile modification | | 4 – 8 | Good for all types of gears. |
| Lapping | | 0 – 10 | Very effective for hypoid gears. |
| Pressure angle | Reduce | 0 – 3 | Reduces tooth stiffness, reduces eccentricity effect, and increases contact ratio. |
| Face width | Increase | | Increases contact ratio for helical gears; reduces deflections |

If all these design and operational changes donot produce adequate noise reduction, we may have to make use of one or all of the following noise control measures [12]:

(a) redesign of gears with higher contact ratio;
(b) adding isolators to the mounts of the gear box;
(c) providing an acoustic hood over the gear box as an extreme solution (as a last resort).

## 6.13 Earthmoving Equipment Noise Estimation and Control

General earthmoving and material handling machinery used in industry generally fall in one of the following groups:

(a) compaction machines like vibration rollers, vibratory plates and vibratory rammers
(b) tracked dozers, tracked loaders and tracked excavator-loaders
(c) wheeled dozers, wheeled loaders, wheeled excavator-loaders, dumpers, graders, loader-type landfill compactors, lift trucks, mobile cranes, nonvibrating rollers and hydraulic power packs
(d) excavators, builders' hoists, construction winches
(e) tower cranes, welding and power generators
(f) compressors
(g) lawnmowers, lawn trimmers and lawn edge trimmers

European Union has specified limits on the permissible A-weighted sound power level of each of the seven groups of machines listed above, as a function of their net installed power (in $kW$), mass of the appliance (in kg), or cutting width (in cm), as applicable.

In particular, compacting machines, including vibrating rollers, and vibratory plates and rammers tend to be the noisiest. Similarly, the handheld concrete breakers and picks are also very noisy because of the vibratory action. Next in that order are tracked dozers, loaders and escalator loaders. A major source of noise in residential localities are lawn mowers and lawn trimmers. In general, wheeled dozers, loaders, dumpers, graders, landfill compactors, combustion engine driven counterbalanced lift trucks and compaction machines are marginally quieter (by a couple of decibels) than their tracked counterparts [13, 14].

A disturbing feature of the earthmoving equipment is that these machines are often used in residential localities and commercial areas. Portable noise barriers must be used to shield the immediate neighbors from the annoyingly excessive noise of these machines.

Legislative noise control has generally proved effective. The legislative limits have been progressively lowered. For example, the stage II (2008) values of $L_{WA}$ are 2 to 3 dB lower then those of stage I (2002). With continued efforts of all machinery manufacturers, the present (2012) levels may be estimated to be lower than those of the stage II values by a couple of decibels.

In fact, legislative control of noise has proved to be effective for all types of industrial machines as well as automobiles and passenger aircraft in reducing their noise decade after decade. The legislative limits have been tightened accordingly, stage after stage, giving sufficient time for the designers as well as manufacturers to design for quietness making use of different practices listed earlier in this chapter.

As indicated before in Section 6.5, noise of stationary compressors (driven by I.C. engines or electric motors) can be contained by means of acoustic hood or enclosure.

In most earthmoving equipment, sources of noise are

(a)  cooling fans
(b)  engines
(c)  electric motors
(d)  gear boxes (transmission systems)
(e)  hydraulic pumps

Noise control measures for all these components have been discussed earlier in this chapter. However, additional measures need to be adopted for the earthmoving equipments on the road or in residential and commercial locations. Portable acoustic barriers are used with limited but considerable success in containing the noise by several decibels at the site of operation like laying of roads, building of metro railway, elevated highways, flyovers, etc.

## 6.14 Impact Noise Control

Material handling operations often involve transfer of material from one level to another. A few of the typical operations are shown in Figs. 6.6 and 6.7.

Impact noise due to metallic components or parts falling into a totebox or stillage can be reduced by

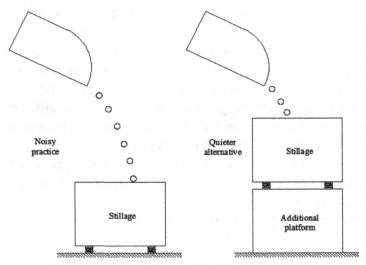

Fig. 6.6  Reducing impact noise, by reducing height.

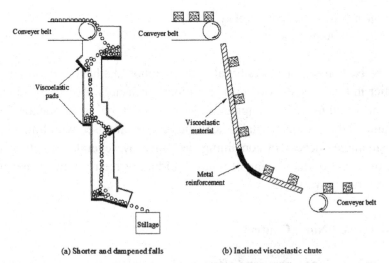

(a) Shorter and dampened falls        (b) Inclined viscoelastic chute

Fig. 6.7  Reducing impact noise, by path modification.

(a)  reducing the height of fall by raising the stillage or totebox onto a
     platform, as shown in Fig. 6.6;
(b)  breaking the free fall into several shorter or milder falls, and
     dampening the impact by means of viscoelastic pads to cushion the
     fall, as shown in Fig. 6.7 (a);

(c) making use of an inclined rubber chute as shown in Fig. 6.7 (b); and making the stillage of Fig. 6.6 out of interwoven metallic strips rather than flat (unperforated) metallic plate.

## 6.15 Environmental Impact Assessment (EIA)

Making use of the empirical expressions for sound power levels of various types of machines given in the last few sections of the current chapter, and the formulae presented in Chapter 4 for acoustics of rooms and enclosures, we can estimate the sound pressure level at a point of interest in the neighborhood (say, at the property line of the industry owner) with and without the noise control measures. This is illustrated below in Example 6.5.

**Example 6.5**  A $2MW$ diesel generator set is proposed to be installed at the edge of a factory at a distance of *20 m* from the property line. Mandatory limits on $L_{pA}$ for industrial area are 75 dB during the day and 70 dB during the night. Conduct an EIA to determine whether these limits would be satisfied without an acoustic enclosure. If not, what should be the minimum insertion loss for which the acoustic enclosure and exhaust muffler need to be designed. The captive power from the DG set is required during the night as well as the day. The diesel engine is turbocharged and the DG set runs at 1500 RPM. The tail pipe length is about 6m, and induction pipe length is 0.9 m.

**Solution**

Use of Eqs. (6.19) – (6.21) for sound power level and the last row of Tables 6.11-6.14 for the A-weighting corrections yields rough estimates of the unmuffled sound power levels as follows.

$$L_{WA}(unmuffled\ exhuast) = 120 + 10\ \log(2 \times 1000) - 6 - 6/1.2 - 12$$

$$= 130\ dB$$

$$L_{WA}(inlet) = 95 + 5\ \log(2 \times 1000) - 0.9/1.8 - 3 = 108\ dB$$

$$L_{WA}(casing) = 94 + 10\ \log(2 \times 1000) - 2 = 125\ dB$$

Assuming that the engine fan noise is included in the casing noise, and that all three sources of noise are nearly equi-distant (*20 m*) from the property line of the factory, the A-weighted SPL at the property line is given by

$$L_{pA}(r) = L_{WA} - 10 \ \log\left(2\pi r^2\right)$$

or

$$L_{pA}(20m) = 130 \oplus 125 \oplus 108 - 10 \ \log\left(2\pi \times 20^2\right)$$
$$= 131 - 34 = 97 \ dB$$

This SPL is $97 - 75 = 22 \ dB$ higher than the day limit, and $97 - 70 = 27 \ dB$ higher than the night-time limit.

Therefore, we need to design the acoustic enclosure and exhaust muffler for insertion loss of at least 27 dB for night-time use of the DG set. However, it is inordinately costlier to design the exhaust muffler as well as acoustic enclosure for IL exceeding 25 dB.

Significantly, however, estimates of Eqs. (6.19) – (6.21) are too conservative. As indicated in the text, the present-day machines are 5-10 dB quieter than their counterparts of three decades ago, when measurements were made for formulation of empirical formulae given in Eqs. (6.19) – (6.21). Therefore, in all probability, the exhaust muffler and acoustic enclosure designed for IL of 25 dB would suffice. Based on this environmental impact assessment (EIA), installation of the DG set may be permitted subject to the factory owners installing acoustic enclosure shown in Fig. 6.8 and a properly designed residential muffler with a volume of at least 10 times the engine capacity or piston displacement (medium or large, as per Table 5.1).

The acoustic enclosure of Fig. 6.8 needs to designed, fabricated and installed as per Section 4.3. In particular the following may be noted:

(a) For acoustic louvers with $l = 1$ m, $2d = 2h = 100$ mm, IL would be about 25 dB.

(b) The outer (impervious) layer of the acoustic enclosure may be a brick wall or at least 1.6 mm thick GI Plate.

(c) For acoustic louvers and lining of the walls, the commercially available 64 kg/m³ density mineral wool or 32 kg/m³ density glass wool may be used. This may be covered with 8 mil (or 10 mil) glass cloth (or tissue) and a thin (26 guage) GI plate or 22 guage aluminium plate with at least 20% porosity.

Features 1 to 10 listed in Section 4.3 must be heeded. Such an acoustic enclosure would yield IL of 25 dBA at least.

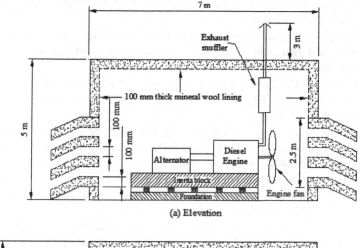

(a) Elevation

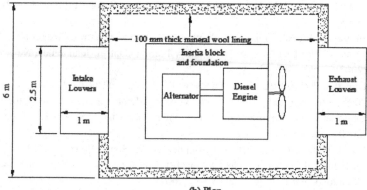

(b) Plan

Fig. 6.8 Schematic views of a DG set with 7 m x 6m x 5 m (high) acoustic enclosure.

# References

1. Bies, D. A. and Hansen, C. H., Engineering Noise Control, Fourth Edition, Spon Press, London, (2009).
2. Joint Departments of the Army, Air Force and Navy, USA, Noise and vibration control for mechanical equipment, Washington, DC: Technical manual TM 5-805-4/AFJMAN 32-1090, (1995).
3. Heitner, I., How to estimate plant noises, Hydrocarbon Processing, 47(2), pp. 67-74, (1968).
4. Edison Electric Institute, Electric power plant environmental noise guide, Bolt Beranek and Newman Inc. Technical Report, (1978).
5. Crocker, M. J., Noise control of compressors, Chapter 74 in M. J. Crocker (Ed.) Handbook of Noise and Vibration Control, John Wiley, New York (2007).
6. Madison, R., Fan Engineering (Handbook), 8$^{th}$ Ed., Buffalo Forge Company, Buffalo, New York, (1938).
7. Graham, J. B. and Hoover, R. M., Fan noise, Chapter 41 in Handbook of Acoustical Measurements and Noise Control, 3$^{rd}$ Ed., C. M. Harris, Ed., McGraw Hill, New York, (1991).
8. Lauchle, C. C., Centrifugal and axial fan noise prediction and control, Chapter 71 in M. J. Crocker (Ed.) Handbook of Noise and Vibration Control, John Wiley, New York, (2007).
9. Cudina, Mirko, Pumps and pumping system noise and vibration prediction and control, Chapter 73 in M. J. Crocker (Ed.) Handbook of Noise and Vibration Control, John Wiley, New York, (2007).
10. Zusman, George, Types of electric motors and noise and vibration prediction and control methods, Chapter 72 in M. J. Crocker (Ed.) Handbook of Noise and Vibration Control, John Wiley, New York, (2007).
11. Baumann, H. D. and Coney, W. B., Noise of gas flows, Chapter 15 in I. L. Ver and L. L. Beranek, (Ed.) Noise and Vibration Control, Second Edition, John Wiley, (2008).
12. Houser, D. R., Gear noise and vibration prediction and control, Chapter 69 in M. J. Crocker (Ed.) Handbook of Noise and Vibration Control, John Wiley, New York, (2007).
13. Directive 2000/14/EC of the European Parliament and of the Council of 8 May 2000 on the Approximation of the Laws of the Member States relating to the Noise Emission in the Environment by Equipment Use Outdoors, Article 12, pp. 6-8, (2000).
14. Bruce, R. D., Moritz, C. T. and Bommer, A. S., Sound Power Level Predictions for Industrial Machinery, Chap. 82 in M. J. Crocker (Ed.), Handbook of Noise and Vibration Control, Wiley, New York, (2007).

**Problems in Chapter 6**

**Problem 6.1** For a 100 kW compressor, evaluate the total exterior sound power level generated, in dB and dBA, for the following cases:

(a) the casing of a rotary (screw type) or reciprocating compressor with a partially muffled inlet,
(b) the casing of a centrifugal compressor, and
(c) the air inlet of a centrifugal compressor,

and thence evaluate the respective A-weighting corrections $\Delta L_{WA}$ to be subtracted from the overall $L_W$ in order to evaluate the overall $L_{WA}$.

[Ans.: (a) 1.0 dB, (b) 2.0 dB, and (c) 0.5 dB]

**Problem 6.2** For a small compressor, making use of the estimated values of $L_p$ (1$m$), evaluate the overall SPL at 1m without and with A-weighting corrections, and thence evaluate the overall A-weighting corrections ($\Delta L_{pA}$) to be subtracted from the overall $L_p$ (1$m$) in order to evaluate $L_{pA}$ (1$m$) for each of the three ranges of compressor power.

[Ans.: $\Delta L_A$ = 1.1 dB for upto 1.5 kW, 1.4 dB for 2 – 6 kW,
and 1.7 dB for 7 – 75 kW]

**Problem 6.3** For the backward-curved centrifugal fan of Example 6.2, estimate the overall linear sound power level, and thence evaluate the overall A-weighting correction ($\Delta L_{WA}$).

[Ans.: $L_W$ = 115.6 dB, and $\Delta L_A$ = 8.2 dB]

**Problem 6.4** For a 30 kW motor running at 1500 RPM, evaluate the overall $L_{pA}$ at 1 m and A-weighting correction $\Delta L_A$, for the case of a

(a) totally enclosed, fan cooled (TEFC) motor, and
(b) drip proof (DRPR) motor

and compare the same with those given in the last row of Table 6.15.

[Ans.: $L_{pA}$ = 89 dB, $\Delta L_A$ = 1 dB, (*b*) $L_{pA}$ = 81 dB, $\Delta L_A$ = 4 dB ]

**Problem 6.5** A pneumatic cleaning convergent nozzle has outlet diameter of 3 mm. The source of compressed air is a receiver with a steady static pressure of 7 bars at 25°C temperature. Estimate the sound pressure level generated by the (highly) choked jet at the operator's ear behind the nozzle at a distance of 0.5 m.

                                                    **[Ans.: 109.4 dB]**

# Nomenclature

Every symbol has been described at the place of its first appearance in the text. Therefore, only those symbols that appear often in the text, are described here.

| | | | |
|---|---|---|---|
| $a$ | Acceleration | $e$ | Eccentricity |
| $A$ | Complex amplitude of the forward progressive wave; Area of cross-section; Total absorption of the room | $E$ | Flow resistivity; Young's Modulus; Electromotive force |
| $AFR$ | Air fuel ratio | $ECTR$ | Extended concentric tube resonator |
| $ANC$ | Active noise control | $E_r$ | Storage modulus |
| $AVC$ | Active vibration control | $E_t$ | Loss modulus |
| $B$ | Number of cylinders of a compressor; Complex amplitude of the rearward progressive wave | $ETEC$ | Extended tube expansion chamber |
| | | $F$ | Froude's friction factor; Force amplitude |
| $BPF$ | Blade passing frequency | $f$ | Frequency (in Hertz); Force |
| $bw$ | Half-power bandwidth | $\{F\}$ | Force vector |
| $bw_n$ | Bandwidth of the n-octave band | $F_0$ | Amplitude of the exciting force |
| $c$ | Speed of sound; Damping coefficient | $f_{dip}$ | Frequency at which a dip or trough occurs in the $TL$ spectrum |
| $c_c$ | Critical damping | | |
| $CLD$ | Constrained layer damping | $F_f$ | Force transmitted to the compliant foundation |
| $CTR$ | Concentric tube resonator | | |
| $D$ | Daily noise dosage; Diffraction directivity factor; Muffler shell diameter | $f_i$ | Lower limit of the frequency band |
| | | $FIR$ | Finite impulse response |
| $d$ | Diameter of the exhaust pipe or tail pipe | $FLD$ | Free layer damping |
| | | $f_m$ | Mean frequency of the band |
| $dB$ | Decibel | $f_{m,n}$ | Natural frequency corresponding to the $(m, n)$ mode of vibration |
| $dB(A)$ | A-weighted decibel | | |
| $dBA$ | A-weighted decibel | | |
| $DF$ | Directivity factor | $f_p$ | Peak frequency |
| $DI$ | Directivity index | $FR$ | Force ratio |
| $DOF$ | Degree of freedom | | |

| | | | |
|---|---|---|---|
| $F_T$ | Force transmitted to the foundation | $L_{pA}$ | A-weighted sound pressure level; Sound level |
| $f_u$ | Upper limit of the frequency band | $L_u$ | Velocity level |
| $G$ | Shear modulus | $L_w$ | Sound power level |
| $g$ | Shear parameter | $m$ | Area ratio; Mass (lumped) |
| $G_2$ | Storage shear modulus of the viscoelastic layer in CLD treatment | $M$ | Mean flow Mach number |
| | | $[M]$ | Inertia matrix |
| $H$ | Dynamic head; specific sound power of a fan | $n$ | Ratio of the area of cross-section of the narrower pipe to that of the wider pipe at a sudden area discontinuity; Speed order |
| $[H]$ | Dynamic matrix | | |
| $h$ | Plate thickness; ratio of flow area to lined perimeter | | |
| $I$ | Acoustic intensity; Second moment of area of cross-section (Moment of Inertia) | $N_{cyl}$ | Number of cylinders |
| | | $N_i$ | Fresnel number in the $i^{th}$ direction |
| $IL$ | Intensity level; Insertion loss | $NR$ | Noise reduction |
| $I_{ref}$ | Reference intensity | $N_s$ | Strouhal number |
| $K$ | Environmental correction; Dynamic pressure loss factor | $n_s$ | Number of surfaces touching at the source |
| | | $N_{st}$ | Number of strokes |
| $k$ | Wave number; Stiffness | $\{O\}$ | Null vector |
| $k_0 l$ | Helmholtz number | $OAF$ | Open area fraction |
| $K_a$ | Acoustical power coefficient | $OAR$ | Open area ratio |
| $k_a$ | Axial stiffness | $P$ | Power of the motor driving a pump; Lined perimeter; Loudness level in phons |
| $k_c$ | Convective wave number | | |
| $k_s$ | Shear stiffness | | |
| $k_x$ | Wave number in the x-direction | $p$ | Acoustic pressure |
| | | $p_0$ | Atmospheric Pressure |
| $k_y$ | Wave number in the y-direction | $p_s$ | Source pressure |
| | | $p_{th}$ | Threshold pressure |
| $k_z$ | Wave number in the z-direction | $Q$ | Directivity factor |
| | | $Q_B$ | Barrier directivity factor |
| $[K]$ | Stiffness matrix | $q_i(t)$ | Modal coordinate for the $i^{th}$ mode |
| $l$ | Length | | |
| $L_a$ | Acceleration level | $Q_l$ | Locational directivity factor |
| $L_d$ | Day-time average SPL; Displacement level | $R$ | Resistance; Real part of the impedance; Gas constant; Reflection coefficient; Non-dimensional acoustic flow resistance; Room constant |
| $L_{dn}$ | Day-night average SPL | | |
| $L_{eq}$ | Equivalent sound pressure level over a period | | |
| $L_I$ | Intensity level | $r$ | Radial coordinate; Radial distance; Frequency ratio |
| $L_n$ | Night-time average SPL | | |
| $l_p$ | Perforate length | $R_2$ | Room constant of the receiver room |
| $L_p$ | Sound pressure level | $RPM$ | Revolutions per minute |

| | | | |
|---|---|---|---|
| $S$ | Area of cross-section; Loudness index in sones; Free-flow area of the cross-section; microphone sensitivity | $\{\ddot{x}\}$ | Acceleration vector |
| | | $Y$ | Characteristic impedance; Amplitude of the displacement excitation |
| $SEC$ | Simple expansion chamber | $y(t)$ | Support displacement |
| $S_m$ | Area of the hypothetical measurement surface | $z$ | Axial coordinate |
| | | $Z$ | Impedance |
| $SPL$ | Sound pressure level | $Z_c$ | Lumped Impedance of the cavity |
| $SSL$ | Source strength level | | |
| $S_w$ | Surface area of the partition wall | $Z_e$ | Lumped impedance of the exhaust pipe |
| $SWL$ | Sound power level | $Z_L$ | Load impedance |
| $T$ | Temperature; Time period | $Z_s$ | Source impedance |
| $t$ | Time variable | $Z_t$ | Lumped Impedance of the tail pipe |
| $[T]$ | Transfer matrix | | |
| $T_{ij}$ | The $i^{th}$ row $j^{th}$ column element of the transfer matrix | $\ominus$ | Logarithmic subtraction |
| | | $\oplus$ | Logarithmic addition |
| $T_{60}$ | Reverberation time | | |
| $TL_a$ | Axial $TL$ | **Subscripts** | |
| $TL_{net}$ | Net $TL$ | | |
| $TL_{tp}$ | Transverse power $TL$ | $a$ | Absorber; Axial; Acoustic |
| $TR$ | Transmissibility | $b$ | Bending |
| $U$ | Mean flow axial velocity | $d$ | Downstream |
| $u$ | Particle velocity | $e$ | Effective |
| $\{u_i\}$ | Modal vector for the $i^{th}$ mode | $eq$ | Equivalent |
| $u_0$ | Initial velocity | $f$ | Flexural |
| $v$ | Volume velocity; Mass velocity; Velocity of a lumped mass | $g$ | Geometric or physical |
| | | $i$ | Incident; The $i^{th}$ mode |
| | | $m$ | Mechanical |
| $V$ | Volume; complex amplitude of velocity | $n$ | Of the neck; Natural |
| | | $r$ | Reflected |
| $W$ | Power flux | $ref$ | Reference value |
| $w$ | Transverse (flexural) displacement | $rms$ | Root mean square |
| | | $t$ | Transmitted |
| $W_{ref}$ | Reference power | $u$ | Upstream |
| $x$ | Open area ratio; Displacement (instantaneous) | | |
| | | **Greek Symbols** | |
| $X$ | Reactance; Imaginary part of the impedance; Displacement amplitude | | |
| | | $\alpha$ | Absorption coefficient of a partition wall |
| $x_0$ | Initial displacement | $\bar{\alpha}$ | Absorption coefficient of an acoustic layer |
| $x_{st}$ | Static displacement or deflection | | |
| $\dot{x}$ | Instantaneous velocity | $\delta$ | End correction; Logarithmic decrement |

$\delta_i$    Difference in the diffracted path and the direct path in the $i^{th}$ direction

$\Delta$    Differential length

$\Delta p$    Stagnation pressure drop; Back pressure

$\gamma$    Ratio of specific heats

$\lambda$    Wave length

$\upsilon$    Poisson's ratio

$\Omega$    Angular speed of rotation

$\omega$    Circular frequency (in radians/second)

$\omega_d$    Damped natural frequency

$\omega_n$    Natural frequency (undamped)

$\rho_0$    Atmospheric density

$\sigma$    Porosity of the perforate

$\tau$    Transmission coefficient

$\xi$    Displacement

$\eta_{G2}$    Loss factor of the viscoelastic material in shear in CLD treatment

$\eta$    Loss factor; Non-dimensional frequency parameter

$\zeta$    Acoustic impedance of a passive subsystem at the upstream end; Damping ratio

# Index